Springer Theses

Recognizing Outstanding Ph.D. Research

For further volumes:
http://www.springer.com/series/8790

Aims and Scope

The series "Springer Theses" brings together a selection of the very best Ph.D. theses from around the world and across the physical sciences. Nominated and endorsed by two recognized specialists, each published volume has been selected for its scientific excellence and the high impact of its contents for the pertinent field of research. For greater accessibility to non-specialists, the published versions include an extended introduction, as well as a foreword by the student's supervisor explaining the special relevance of the work for the field. As a whole, the series will provide a valuable resource both for newcomers to the research fields described, and for other scientists seeking detailed background information on special questions. Finally, it provides an accredited documentation of the valuable contributions made by today's younger generation of scientists.

Theses are accepted into the series by invited nomination only and must fulfill all of the following criteria

- They must be written in good English.
- The topic should fall within the confines of Chemistry, Physics, Earth Sciences, Engineering and related interdisciplinary fields such as Materials, Nanoscience, Chemical Engineering, Complex Systems and Biophysics.
- The work reported in the thesis must represent a significant scientific advance.
- If the thesis includes previously published material, permission to reproduce this must be gained from the respective copyright holder.
- They must have been examined and passed during the 12 months prior to nomination.
- Each thesis should include a foreword by the supervisor outlining the significance of its content.
- The theses should have a clearly defined structure including an introduction accessible to scientists not expert in that particular field.

Keita Fuchise

Design and Precise Synthesis of Thermoresponsive Polyacrylamides

Doctoral Thesis accepted by
Hokkaido University, Sapporo, Japan

Springer

Author
Dr. Keita Fuchise
Faculty of Engineering
Hokkaido University
Sapporo
Japan

Supervisor
Prof. Dr. Toyoji Kakuchi
Faculty of Engineering
Hokkaido University
Sapporo
Japan

ISSN 2190-5053 ISSN 2190-5061 (electronic)
ISBN 978-4-431-56403-4 ISBN 978-4-431-55046-4 (eBook)
DOI 10.1007/978-4-431-55046-4
Springer Tokyo Heidelberg New York Dordrecht London

Softcover reprint of the hardcover 1st edition 2014

Printed on acid-free paper

Springer is part of Springer Science+Business Media (www.springer.com)

Parts of this thesis have been published in the following journal articles

[1] **Fuchise K**, Kakuchi R, Lin S-T, Sakai R, Sato S-i, Chen W-C, Kakuchi T, (2009) Control of Thermoresponsive Property of Urea End-Functionalized Poly(*N*-isopropylacrylamide) Based on the Hydrogen Bond-Assisted Self-Assembly in Water. J Polym Sci, Part A, Polym Chem 47: 6259–6268.

[2] **Fuchise K**, Sakai R, Sato S-i, Satoh T, Narumi A, Kawaguchi S, Kakuchi T, (2010) Group Transfer Polymerization of *N*,*N*-Dimethylacrylamide Using Novel Efficient System Consisting of Dialkylamino Silyl Enol Ether as an Initiator and Strong Brønsted Acid as an Organocatalyst. Macromolecules 43: 5589–5594.

[3] **Fuchise K**, Chen Y, Takada K, Satoh T, Kakuchi T, (2012) Effect of Counter Anions on Kinetics and Stereoregularity for the Strong Brønsted Acid-Promoted Group Transfer Polymerization of *N*,*N*-Dimethylacrylamide. Macromol Chem Phys 213: 1604–1611.

Supervisor's Foreword

Thermoresponsive polymers have been the focus of attention as intelligent material that are capable of reversibly and drastically changing their solubility in certain solvents, such as water and alcohols, depending on the solution temperature. Polyacrylamides are representative thermoresponsive polymers that show a reversible solubility change in water and have been intensively investigated in the past decades. The following are the two key tasks to utilize polyacrylamides as intelligent materials with the desired properties: (1) the development of precise synthesis methods for polyacrylamides with a well-defined primary structure; and (2) elucidation of the relationship between the primary structure and the thermoresponsive properties. Thus, this thesis focused on developing easy and versatile methods to control the thermoresponsive properties of polyacrylamides through designing their primary structure and developing novel precise synthetic methods.

Chapter 1 comprehensively reviews the synthetic methods and thermoresponsive properties of polyacrylamides. Chapter 2 describes the control of the thermoresponsive properties of poly(*N*-isopropylacrylamide) by the introduction of various urea groups with a hydrogen bonding ability to the chain end of the polymer. The precise synthesis of the urea end-functionalized PNIPAMs were achieved by combining the atom transfer radical polymerization (ATRP) and the click reaction. The thermoresponsive properties of the PNIPAM were controlled by tuning the hydrogen-bonding ability of the terminal urea group; the phase-transition temperature decreased with the increasing hydrogen-bonding ability of the terminal urea groups. Chapter 3 describes the development of a novel living polymerization method for acrylamide monomers by the group transfer polymerization (GTP) using a strong Brønsted acid as a precatalyst and an amino silyl enolate as an initiator. The living nature of the GTP of *N*,*N*-dimethylacrylamide (DMAA) was thoroughly investigated. This polymerization was applied to the polymerization of *N*,*N*-diethylacrylamide (DEAA) in Chapter 4 in order to develop a novel and precise synthetic method for poly(*N*,*N*-diethylacrylamide) (PDEAA) with the thermoresponsive properties and its block copolymers, namely thermoresponsive amphiphilic block copolymers and double-hydrophilic block copolymers. The easy and versatile synthesis of various block copolymers bearing a PDEAA segment was achieved by the sequential polymerization of DEAA and DMAA as well as the anionic ring-opening polymerization of epoxides initiated from a hydroxyl end-functionalized PDEAA synthesized by the Tf_2NH-promoted GTP using a functional amino silyl enolate. The thermoresponsive properties of the resultant block copolymers were controlled by the hydrophilicity/hydrophobicity of the introduced polymer.

In summary, the author developed new methods for the precise synthesis of polyacrylamides using the ATRP, the click reaction, and the GTP in order to control the thermoresponsive properties of the polyacrylamides through the design of their primary structure. The result of this research is expected to contribute to the further development of the methodology for designing and synthesizing intelligent materials including thermoresponsive polymers.

Sapporo, September 2013 Prof. Toyoji Kakuchi

Acknowledgments

The research presented in this dissertation was conducted under the direction of Prof. Toyoji Kakuchi, Division of Biotechnology and Macromolecular Chemistry, Faculty of Engineering, Hokkaido University, from 2009 to 2012. I would like to express my deep appreciation to Prof. Kakuchi for his kind instruction, helpful advice, and continuous encouragement during the course of this work.

I am deeply grateful to Associate Prof. Toshifumi Satoh, Division of Biotechnology and Macromolecular Chemistry, Faculty of Engineering, Hokkaido University, and to Associate Prof. Ryosuke Sakai, Department of Materials Chemistry, Asahikawa National College of Technology, for their patient teaching and fruitful daily discussions. I sincerely thank Prof. Seigou Kawaguchi and Associate Prof. Atsushi Narumi, Department of Polymer Science and Engineering, Graduate School of Science and Engineering, Yamagata University, for their helpful support in the size-exclusion chromatography measurements. I wish to express my appreciation to Prof. Wen-Chang Chen and Dr. Sung-Tso Lin, Institute of Polymer Science and Engineering, National Taiwan University, for kindly accepting me into their laboratory and for fine teaching of microscopy. I would like to acknowledge Dr. Harumi Kaga, National Institute of Advanced Industrial Science and Technology (AIST), for his helpful suggestions and advice.

I would also like to give my thanks to all the members of Professor Kakuchi's laboratory for their friendship, accommodating help, and creative, cheerful, and active daily life in the laboratory. In particular, I am grateful to Drs. Yoshikazu Kitajo, Issei Otsuka, Ryohei Kakuchi, Masaki Tamaki, Nguyen To Hoai, and Hideki Misaka for their practical advice and support. I am grateful to Mr. Shigeki Ohta for his fine contribution to this work.

I gratefully acknowledge the Research Fellowships of the Global Center of Excellence (GCOE) program (Catalysis as the Basis for Innovation in Materials Science) and the Grant-in-Aid from the Japan Society for the Promotion of Science (JSPS).

Finally, I would like to express my deep gratitude to my family for their understanding, support, and encouragement throughout this study.

March 2012 Keita Fuchise

Contents

Abbreviations

2VP	2-Vinylpyridine
AMO	*N*-Acryloylmorpholine
APi	*N*-Acryloylpiperidine
APy	*N*-Acryloylpyrrolidine
ATRP	Atom transfer radical polymerization
BO	1,2-Butene oxide
CaH_2	Calcium hydride
CH_2Cl_2	Dichloromethane
CTA	Chain transfer agent
CuAAC	Copper catalyzed azide-alkyne cycloaddition
CuCl	Copper(I) chloride
DEAA	*N*,*N*-Diethylacrylamide
D_h	Hydrodynamic diameter
$\langle D_h \rangle$	Average hydrodynamic diameter
DLS	Dynamic light scattering
DMAA	*N*,*N*-Dimethylacrylamide
DMF	*N*,*N*-Dimethylformamide
DMSO	Dimethylsulfoxide
DP	Degree of polymerization
DPAA	*N*,*N*-Di-*n*-propylacrylamide
DSC	Differential scanning calorimetry
EEGE	1-Ethoxyethyl glycidyl ether
EMAA	*N*-Ethyl-*N*-methylacrylamide
EO	Ethylene oxide
Et_2AlCl	Chlorodiethylalminium
GTP	Group transfer polymerization
HCl	Hydrochloric acid
HPLC	High performance liquid column chromatography
IR	Infrared
K_2CO_3	Potassium carbonate
LCST	Lower critical solution temperature
macro-CTA	Macro chain transfer agent
MALDI-TOF MS	Matrix-assisted laser desorption/ionization time-of-flight mass spectrometry

MALS	Multiangle laser light scattering
Me_6TREN	Tris[2-(dimethylamino)ethyl]amine
MeCN	Acetonitrile
$MgSO_4$	Magnesium sulfate
MMA	Methyl methacrylate
M_n	Number-average molecular weight
$M_{n,calcd}$	Theoretical molecular weight
$M_{n,NMR}$	Number-average molecular weight determined by a ^{1}H NMR measurement
$M_{n,SEC}$	Number-average molecular weight measured by size exclusion chromatography
MOM-NIPAM	*N*-Isopropyl-*N*-methoxymethylacrylamide
MTS	1-Methoxy-1-trimethylsiloxy-2-methyl-1-propene
M_w/M_n	Polydispersity index
MWCO	Molecular weight cut off
Na_2SO_4	Sodium sulfate
NaOH	Sodium hydroxide
NHC	*N*-Heterocyclic carbene
NMP	Nitroxide-mediated polymerization
NMR	Nuclear magnetic resonance
NOESY	Nuclear Overhauser effect correlated spectroscopy
P2VP-*b*-PDEAA	Poly(2-vinylpyridine)-*block*-poly(*N*,*N*-diethyl-acrylamide)
PAA	Poly(acrylamide)
PAA-*b*-PDEAA	Poly(acrylic acid)-*block*-poly(*N*,*N*-diethyl-acrylamide)
PAPy	Poly(*N*-acryloylpyrrolidine)
PBO	Poly(1,2-butene oxide)
PDEAA	Poly(*N*,*N*-diethylacrylamide)
PDEAA-*b*-PBO	Poly(*N*,*N*-diethylacrylamide)-*block*-poly (1,2-butene oxide)
PDEAA-*b*-PEEGE	Poly(*N*,*N*-diethylacrylamide)-*block*-poly (1-ethoxyethyl glycidyl ether)
PDEAA-*b*-PEO	Poly(*N*,*N*-diethylacrylamide)-*block*-poly(ethylene oxide)
PDEAA-*b*-PGD	Poly(*N*,*N*-diethylacrylamide)-*block*-poly(glycidol)
PDEAA-*b*-PPO	Poly(*N*,*N*-diethylacrylamide)-*block*-poly(propylene oxide)
PDEAA-OH	Hydroxyl end-functionalized poly(*N*,*N*-diethyl-acrylamide)
PDMAA	Poly(*N*,*N*-dimethylacrylamide)
PDMAA-*b*-PNIPAM	Poly(*N*,*N*-dimethylacrylamide)-*block*-poly(*N*-isopropylacrylamide)

PDMAA-*b*-PNIPAM-*b*-PDMAA	Poly(*N*,*N*-dimethylacrylamide)-*block*-poly(*N*-isopropylacrylamide)-*block*- poly(*N*,*N*-dimethylacrylamide)
PEEGE	Poly(1-ethoxyethyl glycidyl ether)
PEMAA	Poly(*N*-ethyl-*N*-methylacrylamide)
PEO	Poly(ethylene oxide)
PGD	Poly(glycidol)
PMAA	Poly(*N*-methylacrylamide)
PMDETA	*N*,*N*,*N′*,*N″*,*N″*-Pentamethyldiethylenetriamine
P*n*BMAA	Poly(*N*-*n*-butyl-*N*-methylacrylamide)
PNIPAM	Poly(*N*-isopropylacrylamide)
PNIPAM-*b*-P4VP	Poly(*N*-isopropylacrylamide)-*block*-poly (4-vinylpyridine)
PNIPAM-*b*-PTHP	Poly(*N*-isopropylacrylamide)-*block*-poly (2-(tetrahydropyranyloxy)ethyl methacrylate)
P*n*PAA	Poly(*N*-*n*-propylacrylamide)
PO	Propylene oxide
PPhAA	Poly(*N*-phenylacrylamide)
PPO	Poly(propylene oxide)
P*t*BAA	Poly(*N*-*tert*-butylacrylamide)
PTBTP	Pentaerythritol tetrakis(3-(benzylthiothiocarbonylthio)propionate
RAFT	Reversible addition-fragmentation chain transfer
RI	Refractive index
ROP	Ring-opening polymerization
SEC	Size exclusion chromatography
TAS-HF_2	Tris(dimethylamino)sulfonium hydrogen bifluoride
TAS-$SiMe_3F_2$	Tris(dimethylamino)sulfonium difluorotrimethylsilicate
*t*BA	*tert*-Butyl acrylate
TBA-AcO	Tetra-*n*-butylammonium acetate
TBAF	Tetra-*n*-butylammonium fluoride
*t*BMA	*tert*-butyl methacrylate
*t*BuOH	*tert*-butyl alcohol
t-Bu-P_4	1-*tert*-Butyl-4,4,4-tris(dimethylamino)-2,2-bis-[tris(dimethylamino)-phosphoranylidenamino]-2Λ^5,4Λ^5-catenadi(phosphazene)
T_C	Critical temperature of the phase transition
TEM	Transmission electron microscopy
Tf_2NH	Bis(trifluoromethanesulfonyl)imide
Tf_2NSiMe_3	*N*-(Trimethylsilyl)bis(trifluoromethanesulfonyl) imide
T_g	Glass transition temperature

THF	Tetrahydrofuran
TIPNO	2,2,5-Trimethyl-4-phenyl-3-azahexane-3-nitroxide
TMEDA	*N*,*N*,*N*′,*N*′-Tetramethylethylenediamine
UCST	Upper critical solution temperature
UV-vis	Ultraviolet-visible
(*Z*)-DATP	(*Z*)-1-Dimethylamino-1-trimethylsiloxy-1-propene
(*Z*)-HDATP	(*Z*)-1-(*N*-Methyl-*N*-(2-trimethylsiloxyethyl) amino)-1-trimethylsiloxy-1-propene
ZnI_2	Zinc(II) iodide

Chapter 1
General Introduction

1.1 Thermoresponsive Polymers

Thermoresponsive polymers have been attracting much attention over the past few decades as a promising component of intelligent materials for applications in various fields, such as bioengineering [1–4] and nanotechnology [5–8]. Such polymers exhibit a volume phase transition by a sudden change in the solvation state observed as a coil-to-globule transition of the polymer structure at a critical temperature, such as the lower critical solution temperature (LCST) and the upper critical solution temperature (UCST). The LCST is defined as the temperature at which a thermoresponsive polymer reversibly changes its properties from water-soluble to water-insoluble upon heating, whereas the UCST is when the polymer goes from hydrophobic to hydrophilic, as depicted in Fig. 1.1.

Scheme 1.1 shows the representative thermoresponsive polymers with an LCST or an UCST. Various polyacrylamides [9, 10], poly[oligo(ethylene glycol) methacrylate]s [11–13], poly(vinyl ether)s [14–17], poly(glycidyl ether)s [18–20], poly(2-oxazoline)s [21], and poly(*N*-vinyllactam)s [22–24] have been revealed to have an LCST. Poly(sulfobetaine)s are the representative thermoresponsive polymers with an UCST [25].

In addition to the homopolymer of thermoresponsive polymers, block copolymers consisting of a thermoresponsive polymer and another polymer have attracted much attention because they can reversibly form a micellar assembly in an aqueous solution, of which the behavior is expected to be applied to the non-covalent encapsulation and controlled release of guest molecules for a drug delivery system and other applications [26–32]. A thermoresponsive amphiphilic copolymer, namely a block copolymer consisting of a thermoresponsive polymer and a water-insoluble polymer, forms micellar assemblies, such as polymeric micelles and vesicles, in an aqueous solution when the thermoresponsive segment is water-soluble, while it precipitates or forms larger aggregates upon a temperature change. A double-hydrophilic polymer, such as a block copolymer consisting of a thermoresponsive polymer and a water-soluble polymer, is known

K. Fuchise, *Design and Precise Synthesis of Thermoresponsive Polyacrylamides*, Springer Theses, DOI: 10.1007/978-4-431-55046-4_1, © Springer Japan 2014

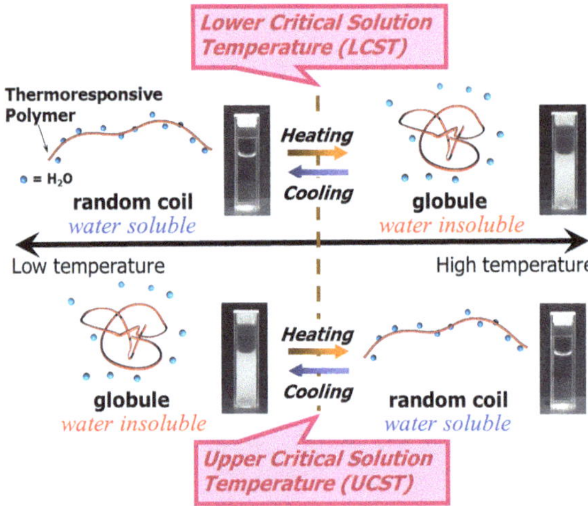

Fig. 1.1 Schematic representation of the phase transition behavior of a thermoresponsive polymer with a lower critical solution temperature (*LCST*) and an upper critical solution temperature (*UCST*)

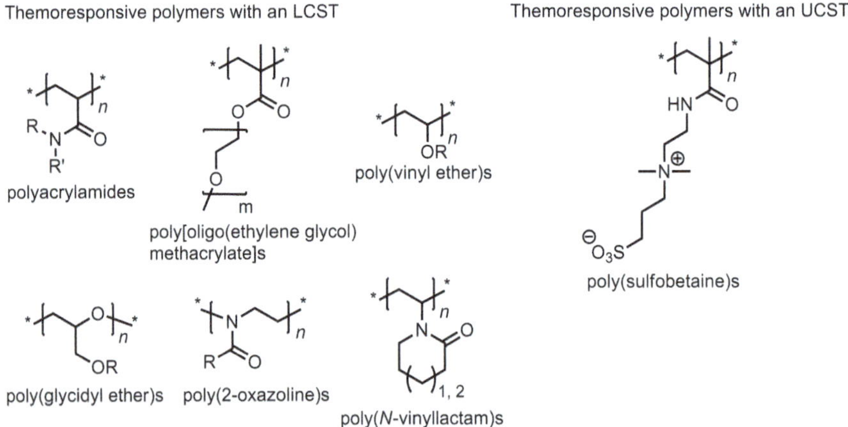

Scheme 1.1 Representative thermoresponsive polymers with an LCST and an UCST

to reversibly form a micellar assembly in an aqueous solution depending on the solution temperature. Figure 1.2 depicts the reversible aggregation behavior of a thermoresponsive amphiphilic block copolymer and a double-hydrophilic block copolymer bearing a thermoresponsive segment with an LCST. The aggregation behavior is inverted when the utilized thermoresponsive polymer is with an UCST. Furthermore, thermoresponsive block copolymers with a more complicating structure, such as a double-hydrophilic block copolymer consisting of a thermoresponsive segment with an LCST and another thermoresponsive segment with an UCST [25] and a triblock copolymer bearing one or two thermoresponsive segment(s)

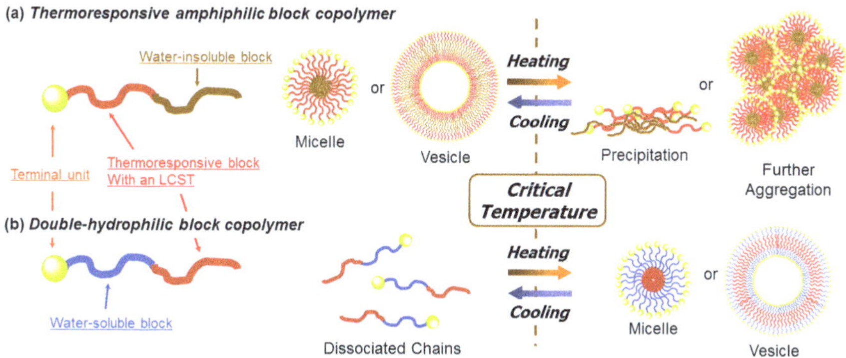

Fig. 1.2 Schematic representation of the reversible formation of micellar assemblies from a thermoresponsive amphiphilic block copolymer and a double-hydrophilic block copolymer

[33, 34], were also synthesized to reversibly control not only the formation of the assembly, but also the size and shape of the resultant assembly. The rational design and the precise synthesis are required to obtain thermoresponsive block copolymers with the desired properties.

1.2 Properties of Polyacrylamides

Among the thermoresponsive polymers, the polyacrylamides have been intensively investigated due to their ease of synthesis and molecular design. The solubility and thermoresponsive properties of polyacrylamides in water vary depending on the hydrophobicity of the substituent groups on their amide groups, as shown in Scheme 1.2. For example, poly(acrylamide) (PAA), poly(*N*-methylacrylamide) (PMAA), and poly(*N*,*N*-dimethylacrylamide) (PDMAA) are water-soluble, while polyacrylamides bearing hydrophobic substituents, such as poly(*N*-*tert*-butylacrylamide) (P*t*BAA), poly(*N*-*n*-butyl-*N*-methylacrylamide) (P*n*BMAA), and poly(*N*-phenylacrylamide) (PPhAA), are water-insoluble. Polyacrylamides bearing moderately hydrophobic substituents, such as poly(*N*-isopropylacrylamide) (PNIPAM), poly(*N*,*N*-diethylacrylamide) (PDEAA), poly(*N*-*n*-propylacrylamide) (P*n*PAA), poly(*N*-ethyl-*N*-methylacrylamide) (PEMAA), and poly(*N*-acryloylpyrrolidine) (PAPy), have a thermoresponsive properties and show LCSTs at 32, 32, 25, 70, and 50 °C, respectively. Thus, the design of the substituent groups has enabled to control the solubility and thermoresponsive properties of the polyacrylamides.

The primary structure of the polyacrylamides, such as molecular weight [35, 36], stereoregularity [37–48], chain-end structure [36, 49–51], branching structure [52, 53], and topology [54–57], is also known to affect the solubility and the LCST in water. Thus, the precise synthesis of polyacrylamides with predetermined molecular weights, narrow molecular weight distributions, and well-defined

Scheme 1.2 Representative polyacrylamides and the difference in their solubility and thermoresponsive properties in water

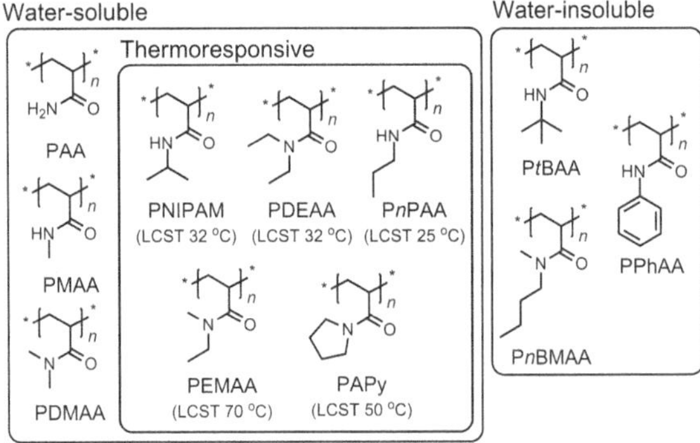

structures has been an important subject not only for elucidating the relationship between the structure and the properties of the polyacrylamides, but also for preparing intelligent materials with the desired properties.

1.3 Historical Aspect of Precise Synthesis for Polyacrylamides

The rapid development of controlled/living polymerizations, such as anionic polymerization, coordination polymerization, nitroxide-mediated polymerization (NMP) [58], atom transfer radical polymerization (ATRP) [59, 60], and reversible addition-fragmentation chain transfer (RAFT) polymerization [61, 62], for acrylamide monomers was the breakthrough in the chemistry of polyacrylamides, which came in the late-1990s to early-2000s and provided methods to control the molecular weight and chain-end structure of the resulting polyacrylamides.

In 1996, Hogen-Esch and Xie first achieved the living polymerization of an acrylamide monomer [37]; the polymerization of *N*,*N*-dimethylacrylamide (DMAA) was carried out using triphenylmethylcesium as the initiator in tetrahydrofuran (THF) at −78 °C to obtain PDMAA having a predetermined molecular weight and narrow molecular weight distribution. Nakahama et al. succeeded in the living anionic polymerization of DMAA, *N*,*N*-diethylacrylamide (DEAA), *N*-acryloylpyrrolidine (APy), *N*-acryloylpiperidine (APi), *N*,*N*-di-*n*-propylacrylamide (DPAA), *N*-ethyl-*N*-methylacrylamide (EMAA), and *N*-acryloylmorpholine (AMO) using diphenyllithium or diphenylpotassium derivatives as initiators in the presence of a Lewis acidic additive, such as diethylzinc(II) and triethylborane, which expanded the scope of the acrylamide monomers that can be polymerized in a living fashion by the anionic polymerization [38, 39]. In addition, Ishizone et al. achieved the precise synthesis of PNIPAM by the polymerization of *N*-isopropyl-*N*-methoxymethylacrylamide (MOM-NIPAM), a NIPAM protected by a methoxymethyl group, in THF at –78 °C using diphenylpotassium and diethylzinc(II) [40, 41]. Scheme 1.3 summarizes the living anionic polymerization of the acrylamides.

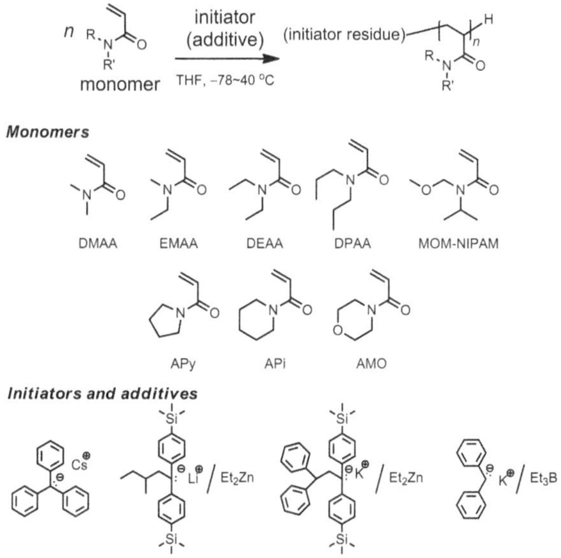

Scheme 1.3 Living anionic polymerization of various acrylamides

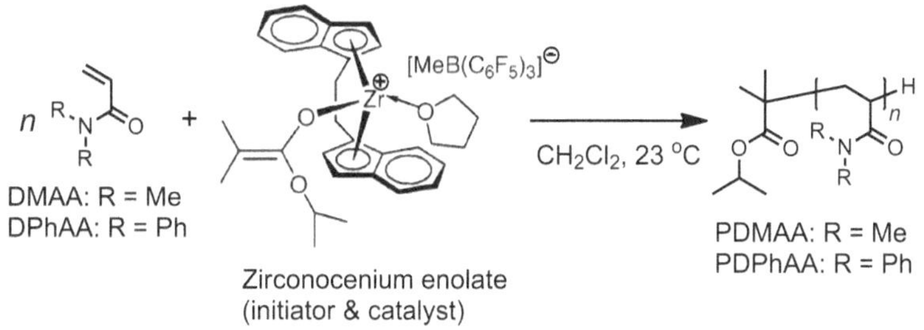

Scheme 1.4 Living coordination polymerization of acrylamide monomers

Chen and Mariott achieved the coordination polymerization of acrylamide monomers initiated and catalyzed by zirconocenium enolate, as shown in Scheme 1.4 [42], which produced the PDMAA [42, 43] and poly(*N*,*N*-diphenylacrylamide) [44] with predetermined molecular weights, extremely narrow molecular weight distributions, and very high isotacticity.

Both the anionic polymerization and the coordination polymerization required strict removal of impurities from the polymerization system and a special apparatus for the polymerization, though they were capable of providing excellent control of the polymerization. Thus, the controlled/living radical polymerizations [63] for acrylamides have been intensively investigated since the discovery of the NMP, the ATRP, and the RAFT polymerization because the ease of polymerization and the tolerance to the presence of impurities. Hawker et al. reported the first successful controlled/living radical polymerization of an acrylamide monomer in 1999, which was achieved by the NMP of DMAA using the specially designed nitroxide,

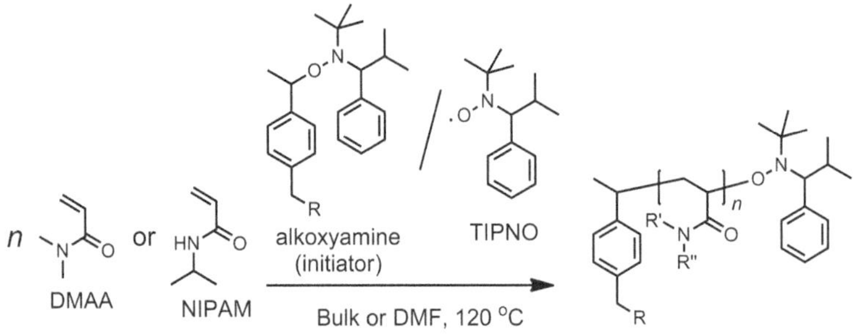

Scheme 1.5 Nitroxide-mediated radical polymerization (*NMP*) of acrylamide monomers

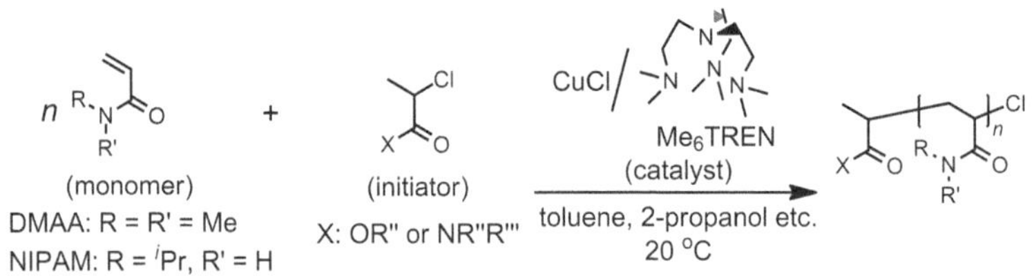

Scheme 1.6 Atom transfer polymerization (*ATRP*) of acrylamide monomers

2,2,5-trimethyl-4-phenyl-3-azahexane-3-nitroxide (TIPNO), and the corresponding alkoxyamine under bulk condition at 120 °C, as shown in Scheme 1.5 [64]. Binder et al. reported the precise synthesis of the end-functionalized PNIPAM using the functionalized alkoxyamine having a TIPNO skeleton in 2007 [65].

ATRP is more advantageous than the other controlled/living radical polymerizations in terms of the ease of designing an initiator, though the complete removal of the utilized metallic catalyst from the product is cumbersome. The most versatile system for the ATRP of acrylamide monomers is that using 2-chloropropionate or 2-chloropropionamide derivatives as the initiator and the complex of copper(I) chloride (CuCl) and tris[2-(dimethylamino)ethyl]amine (Me_6TREN) as the catalyst, as shown in Scheme 1.6. Matyjaszewski et al. reported the first controlled polymerization of DMAA by ATRP using a complex of CuCl and Me_6TREN in toluene at 20 °C in 2000 [66, 67]. Various polymer architectures consisting of thermoresponsive polyacrylamides, especially PNIPAM, have been synthesized using ATRP since Masci et al. [68] and Xia et al. [35, 49] independently reported the controlled polymerization of NIPAM by the ATRP using the complex of CuCl and Me_6TREN.

The RAFT polymerization is currently considered to be the most versatile controlled/living radical polymerization method due to its broad scope of polymerizable monomers and its metal-free nature. The controlled polymerizations of various acrylamide monomers were achieved by the RAFT polymerization using dithiobenzoate and trithiocarbonate derivatives as the chain transfer agent (CTA) in various solvents, such as water, dimethylsulfoxide (DMSO), and 1,4-dioxane, as shown in Scheme 1.7 [69–72]. Gilbert et al. reported the polymerization of NIPAM using

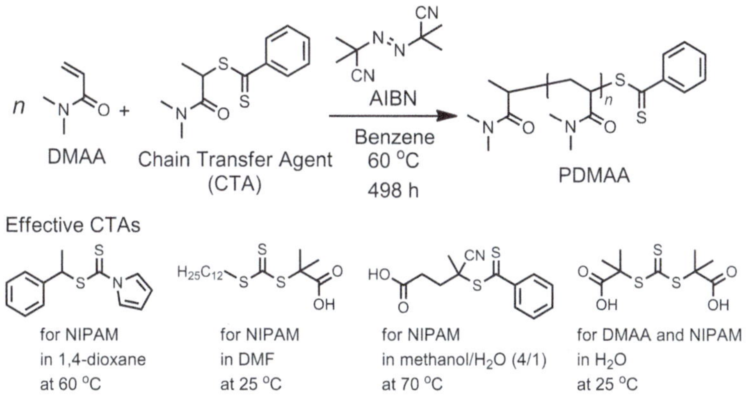

Scheme 1.7 Reversible addition-fragmentation chain transfer (*RAFT*) radical polymerization of acrylamides

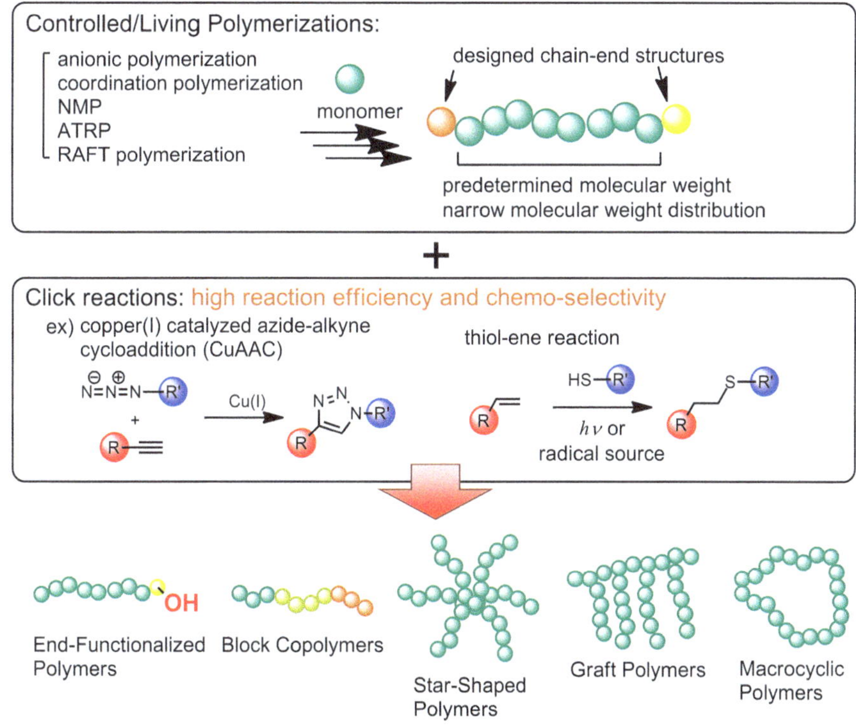

Fig. 1.3 Well-defined polymers synthesized by the controlled/living polymerizations and the click reactions

benzyl dithiobenzoate as the CTA in benzene at 60 °C to produce PNIPAM with a polydispersity index (M_w/M_n) in the range of 1.1–1.5, which was the first controlled polymerization of acrylamide monomers by the RAFT polymerization [73].

Not only the controlled/living polymerization methods but also the rapid development of the click reactions, which features high reaction efficiency and chemo-selectivity, have provided the efficient synthetic method for polyacrylamides with well-defined structure including end-functionalized polymers, block copolymers, star-shaped polymers, macrocyclic polymers, graft copolymers, etc., as shown in Fig. 1.3 [74–78]. Nowadays, many reactions, such as the copper catalyzed azide-alkyne cycloaddition (CuAAC) and thiol-ene reaction, are categorized as the click reaction and utilized for the post-modifications of polymers, even though some of the reactions do not meet the requirements for the click reaction originally defined by Sharpless et al.

1.4 Properties of Well-Defined Thermoresponsive Polyacrylamides and Their Block Copolymers

The precisely synthesized polyacrylamides have begun to provide insights into the influence of the primary structure of polyacrylamides on the solubility and thermoresponsive properties. Winnik et al., Stöver et al., and Kakuchi et al. have intensively investigated the effect of the molecular weight and chain end structure on the thermoresponsive properties of the well-defined PNIPAM synthesized by the ATRP or the RAFT polymerization. Winnik et al. synthesized α,ω-dioctadecyl-PNIPAM with the number-average molecular weight (M_n) that ranged from 12.0 to 49.0 kg mol^{-1} and M_w/M_n lower than 1.20 by the RAFT polymerization using a CTA bearing two octadecyl groups, as shown in Scheme 1.8a. The critical temperature of the synthesized PNIPAM increased from 25 to 31 °C with the increasing molecular weight when the polymer concentration in an aqueous solution was 1.0 g L^{-1} [36]. Stöver et al. synthesized PNIPAM bearing five different chain-end structures with varying hydrophobicities by the ATRP using designed initiators, as shown in Scheme 1.8b [49]. Kakuchi et al. synthesized a series of well-defined PNIPAMs bearing different chain-end groups with varying hydrophobicities by the ATRP, as shown in Scheme 1.8c [50, 51]. In all the results, the hydrophilic end groups led to an increase in the critical temperature, while the hydrophobic end groups decreased it. The influence of the chain end structure on the critical temperature decreased as the molecular weight of the polymer increased.

Star-shaped and hyperbranched polyacrylamides have been synthesized to elucidate the effect of the branching structure on the thermoresponsive properties. Whittaker et al. synthesized the 4-armed star-shaped PNIPAM by the RAFT polymerization of NIPAM using a star-shaped CTA, pentaerythritol tetrakis(3-(benzylthiothiocarbonylthio)propionate (PTBTP), bearing four trithiocarbonate groups, as shown in Scheme 1.9a [56]. The critical temperature of the resultant polymer was lower than the linear PNIPAMs cleaved from the star-shaped PNIPAM most likely due to the hydrophobicity of the core unit and the terminal benzyl groups. The critical temperature increased with the increasing molecular weight of the star-shaped PNIPAM because the influence of the terminal structure on the

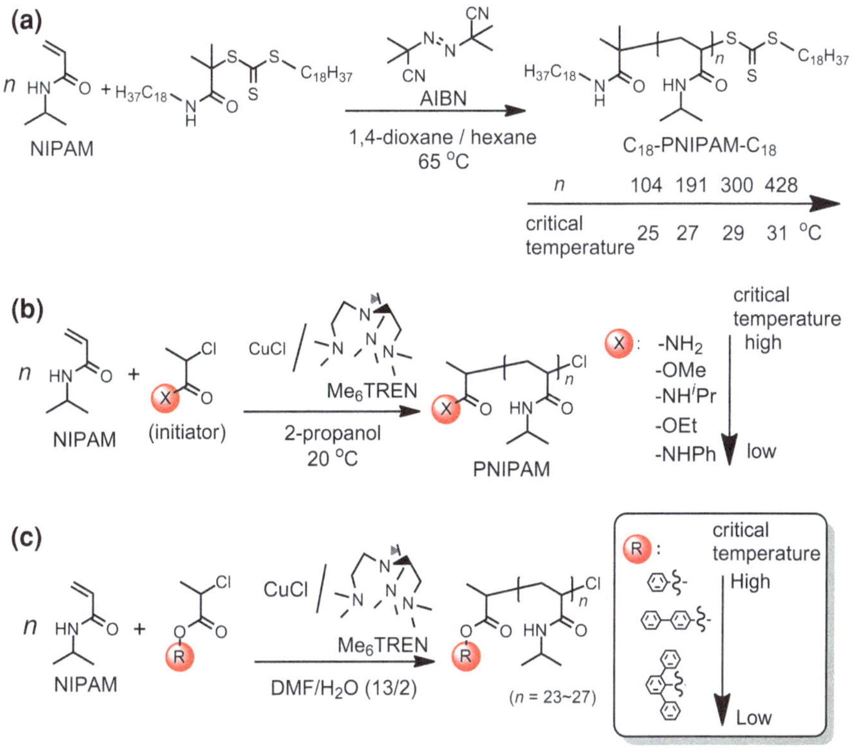

Scheme 1.8 Precise synthesis of PNIPAM for investigating the effect of the molecular weight and chain-end structure on its thermoresponsive properties

thermoresponsive properties decreases as the polymer chain becomes longer. Liu et al. synthesized 7-armed and 21-armed star-shaped PNIPAMs by the CuAAC between the ethynyl end-functionalized PNIPAM and the β-cyclodextrin functionalized with 7 or 21 azido groups, as shown in Scheme 1.9b [79]. The critical temperature of the resultant star-shaped PNIPAM increased with the increasing molecular weight and increasing number of arm polymers. Both Whittaker et al. and Liu et al. attributed the decrease in the critical temperature of the star-shaped PNIPAM with a low molecular weight to the ease of cluster formation by the monomeric units located adjacent to the core unit due to the high density of the grafting chains. Sumerlin et al. synthesized the hyperbranched PNIPAM by the RAFT polymerization of NIPAM using a CTA bearing a trithiocarbonate group and an acryloyl group, as shown in Scheme 1.10 [57]. The critical temperature of the resultant PNIPAM significantly decreased as compared to a linear PNIPAM because the contribution of the hydrophobic chain-end groups, namely the *n*-dodecyl groups, on the thermoresponsive properties increased as the number of chain-ends increased.

Macrocyclic polyacrylamides have been synthesized with PNIPAM. Winnik et al. synthesized the macrocyclic PNIPAM through the synthesis of α-azido-ω-ethynyl PNIPAM by the RAFT polymerization and the macrocyclization reaction

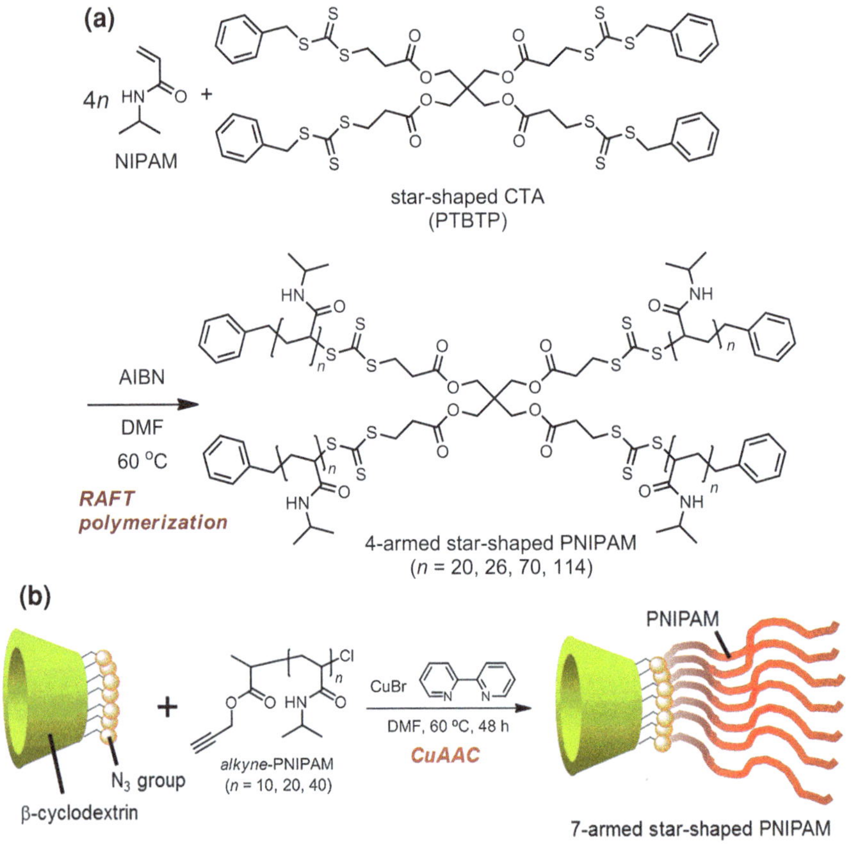

Scheme 1.9 Precise synthesis of star-shaped PNIPAMs for investigation of the relationship between the branching structure and the thermoresponsive properties

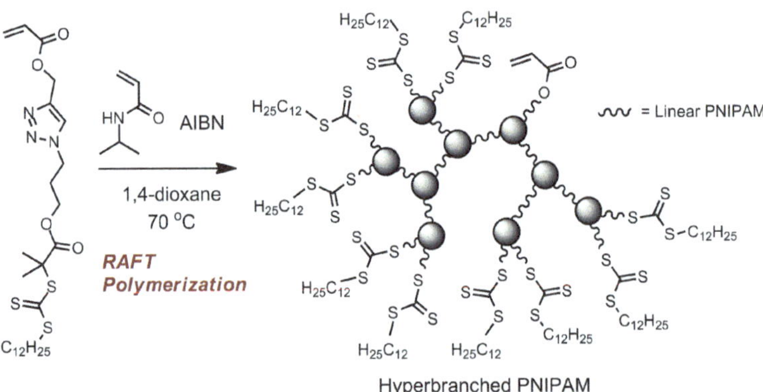

Scheme 1.10 Precise synthesis of the hyperbranched PNIPAM

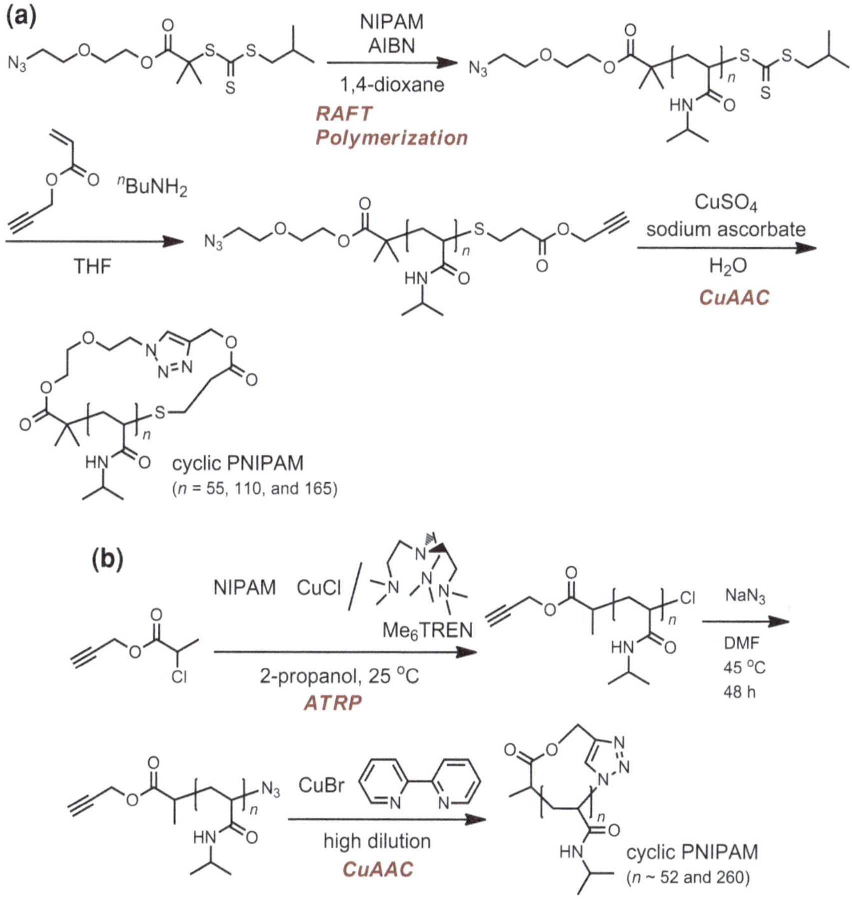

Scheme 1.11 Precise synthesis of macrocyclic PNIPAM

using the CuAAC, as shown in Scheme 1.11a [52, 55]. The critical temperature of the obtained cyclic PNIPAM was higher than that of the linear precursor regardless of the molecular weight and decreased with the increasing molecular weight as with the linear precursors. However, the aggregation process of the cyclic PNIPAM was slower than the linear precursors presumably because the absence of chain-ends hindered the packing of the polymer chains upon the coli-to-globule transition. Liu et al. synthesized the macrocyclic PNIPAM by the combination of the ATRP and the CuAAC, as shown in Scheme 1.11b [53, 54]. In their case, the critical temperature of the resultant cyclic PNIPAM was lower than the linear precursor, though those for both of them decreased with the increasing molecular weight. As a result of laser light scattering and stopped-flow temperature-jump measurements, they revealed that the cyclic PNIPAM tends to form smaller and stable mesoglobules with a relatively lower chain density presumably due to the lack of any interchain entanglement and penetration.

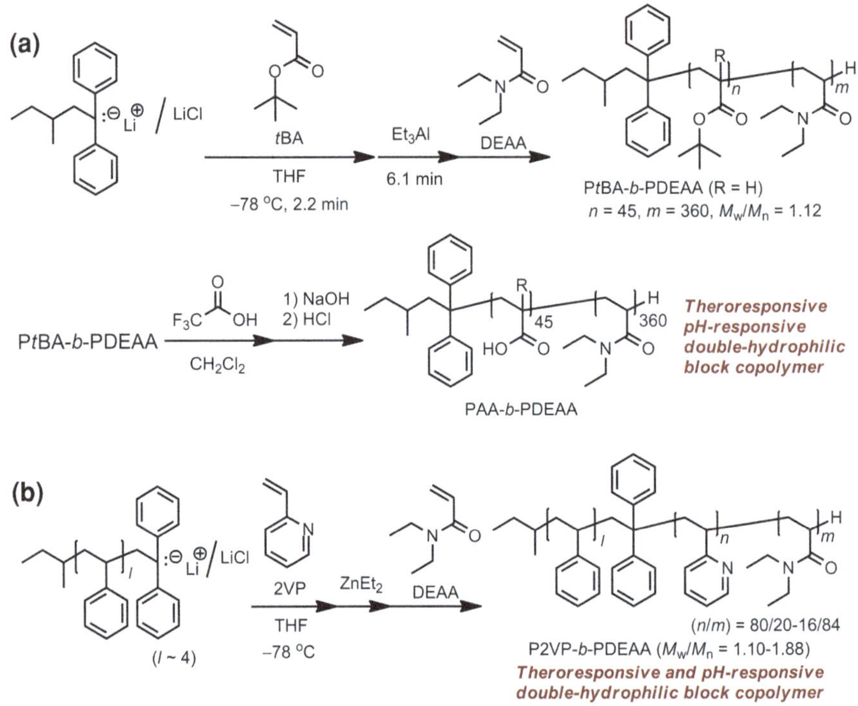

Scheme 1.12 Precise synthesis of double-hydrophilic block copolymers bearing a thermoresponsive polyacrylamide segment by the sequential polymerization

The precise synthesis of polyacrylamides has applied to the synthesis of thermoresponsive amphiphilic block copolymers and double-hydrophilic block copolymers bearing one or multiple thermoresponsive polyacrylamide segment(s) with following synthetic strategies: (1) sequential polymerization of multiple monomers, (2) a polymerization using a macroinitiator or a macro chain transfer agent (macro-CTA), and (3) the polymer-polymer conjugation reaction.

The first approach requires that the polymerization can be continued until the monomer is quantitatively consumed; the addition of the second monomer before the completion of the first polymerization causes the formation of the partially gradient structure in the sequence of monomer units. Thus, only the anionic polymerization has been utilized to synthesize block copolymers of thermoresponsive polyacrylamides with this approach because the controlled/living radical polymerizations should be quenched before the quantitative consumption of the monomer in order to avoid any side reactions, such as the radical-radical coupling reaction. For example, Müller et al. reported the synthesis of poly(acrylic acid)-*block*-PDEAA (PAA-*b*-PDEAA) by the sequential anionic polymerization of *tert*-butyl acrylate and DEAA followed by the deprotection of the *tert*-butyl group from the resultant block copolymer, as shown in Scheme 1.12a [80]. The ability of PAA-*b*-PDEAA to reversibly form micellar aggregates was proved by

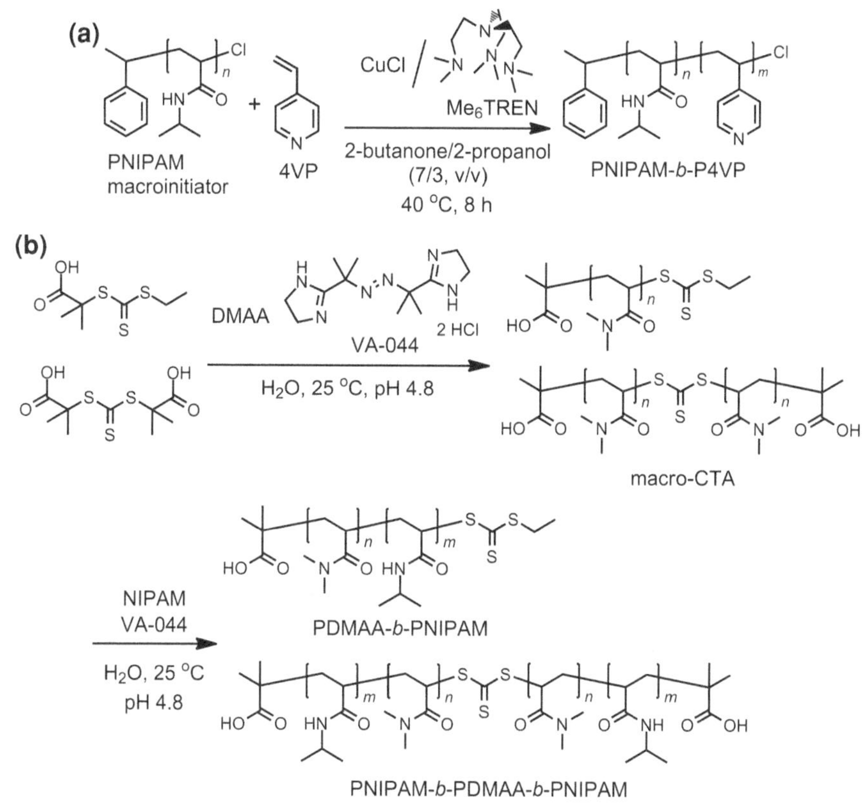

Scheme 1.13 Precise synthesis of double-hydrophilic block copolymers bearing a thermoresponsive polyacrylamide segment by the polymerization using a macroinitiator or a macro-CTA

dynamic light scattering measurements in acidic and alkaline aqueous solutions at different temperatures. Adler et al. reported the synthesis of poly(2-vinylpyridine)-*block*-PDEAA (P2VP-*b*-PDEAA) by the sequential anionic polymerization of 2-vinylpyridine (2VP) and DEAA using the alkyllithium derived from *sec*-butyllithium, styrene, and diphenyl ethylene as the initiator and lithium chloride and diethylzinc as the additives, as shown in Scheme 1.12b [81]. The resultant P2VP-*b*-PDEAA reversibly formed a micellar assembly in an acidic aqueous solution at pH 2 upon heating most likely due to the formation of the hydrophobic core from the dehydrated PDEAA segment.

The second approach, namely the polymerization using a macroinitiator or a macro-CTA, has been applied to the synthesis of various thermoresponsive block copolymers bearing polyacrylamide segments. Shi et al. synthesized PNIPAM-*block*-poly(4-vinylpyridine) (PNIPAM-*b*-P4VP) by the ATRP of 4-vinylpyridine using a macroinitiator composed of PNIPAM, as shown in Scheme 1.13a [82]. The resultant polymer was capable of reversibly forming a micellar assembly in an aqueous solution depending on the temperature and pH. McCormick et al.

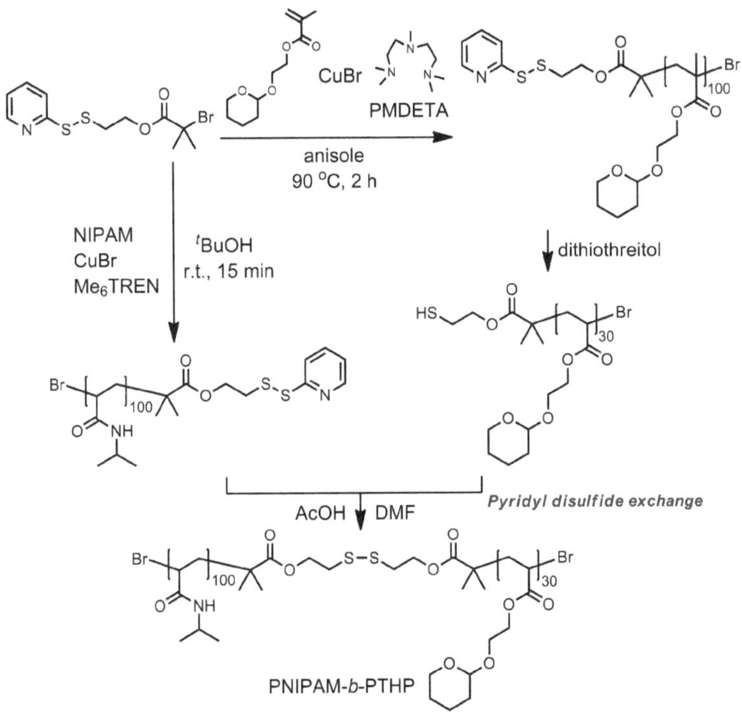

Scheme 1.14 An example of the precise synthesis of thermoresponsive amphiphilic block copolymer bearing a thermoresponsive polyacrylamide segment by the polymer-polymer conjugation reaction

synthesized double-hydrophilic block copolymers, such as PDMAA-*b*-PNIPAM and PDMAA-*b*-PNIPAM-*b*-PDMAA, by the RAFT polymerization of NIPAM using a PDMAA with one or two trithiocarbonate group(s) as the macro-CTA, as shown in Scheme 1.13b [72]. Both of the obtained polymers have an ability to reversibly form a micellar assembly in an aqueous solution. The size of the micellar assembly increased and the critical temperature for the micelle formation decreased as the PNIPAM segment became longer.

The third approach, the polymer-polymer conjugation reaction, has also been utilized to synthesize amphiphilic thermoresponsive block copolymers possessing a polyacrylamide segment. Thayumanavan et al. reported the synthesis of PNIPAM-*block*-poly(2-(tetrahydropyranyloxy)ethyl methacrylate) (PNIPAM-*b*-PTHP) by the pyridyl disulfide exchange reaction between a 2-pyridyldithio end-functionalized PNIPAM and a thiol end-functionalized poly(2-(tetrahydropyranyloxy)ethyl methacrylate), as shown in Scheme 1.14 [83]. The resultant PNIPAM-*b*-PTHP was capable of forming a micellar structure in an aqueous solution and encapsulating hydrophobic guests. The micelle structure can be degraded by heating or acidifying the solution as well as adding reductants to the solution, which was utilized for the controlled release of the encapsulated guests.

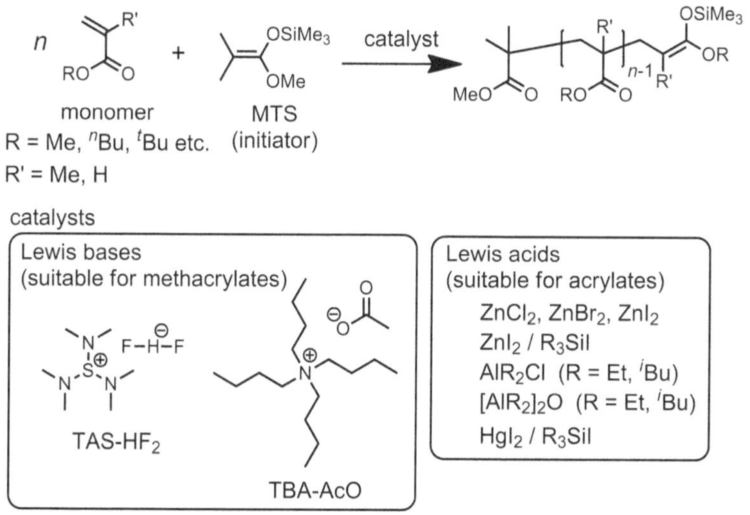

Scheme 1.15 General scheme of the group transfer polymerization (*GTP*)

1.5 Potential of Group Transfer Polymerization for the Synthesis of Well-Defined Polyacrylamides

Group transfer polymerization (GTP) discovered by Webster et al. in 1983 is one of the living polymerization methods for methacrylates and acrylates, which proceeds through the repetitive Mukaiyama-Michael reaction catalyzed by a Lewis base or a Lewis acid, as shown in Scheme 1.15 [84–86]. A silyl enolate, such as 1-methoxy-1-trimethylsiloxy-2-methyl-1-propene (MTS), is generally used as the initiator. Lewis bases, such as the tris(dialkylamino)sulfonium salts and the tetraalkylammonium salts of $SiMe_3F_2^-$ [84, 87, 88], HF_2^- [84, 85, 87, 88], F^- [85, 87, 89], CN^- [84, 85, 87, 89], N_3^- [84, 85], oxyanions [90], and hydrogen bioxyanions [90, 91], have been recognized as suitable catalysts only for the methacrylates. Their catalytic activities are too high to control the polymerization of acrylates that have a higher reactivity than the methacrylates. Lewis acids, such as zinc halides [85, 92], organoaluminiums [85, 92], and mercury(II) iodide [93–96], have been reported to be effective only for acrylates because of their low catalytic activity.

Advantages of the GTP in comparison to other controlled/living polymerizations are as follows: (1) the polymerization is principally free from termination reactions thus realizing excellent control of the polymerization as with anionic polymerization; (2) the initiator can be easily designed for the synthesis of an end-functionalized polymer as with the ATRP; and (3) the polymerization can be carried out at room temperature without using metallic reagents as with the RAFT polymerization. However, the GTP of acrylamide monomers has not been well

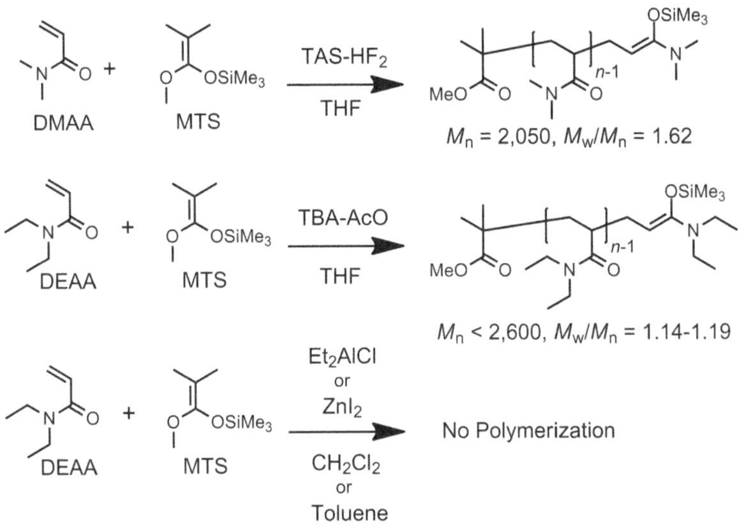

Scheme 1.16 Previously reported GTP of acrylamide monomers

developed, though there have been some attempts to apply the GTP to acrylamide monomers, as shown in Scheme 1.16. For example, Sogah et al. obtained PDMAA with a 2 kg mol^{-1} number average molecular weight (M_n) and a relatively broad M_w/M_n (=1.62) as a result of the GTP of DMAA using MTS as the initiator and tris(dimethylamino)sulfonium hydrogen bifluoride (TAS-HF_2) as the catalyst [85]. Freitag et al. attempted the GTP of DEAA using MTS as the initiator and tetra-*n*-butylammonium acetate (TBA-AcO), chlorodiethylaluminium (Et_2AlCl), or zinc(II) iodide (ZnI_2) as the catalyst. The polymerization using TBA-AcO produced PDEAA with smaller than 3 kg mol^{-1} of M_n of and 1.14–1.19 of M_w/M_n, while those using Et_2AlCl or ZnI_2 did not proceed [97–99].

Recently, a new class of catalysts, so-called organocatalysts, has been applied to the GTP to produce polymers with a controlled molecular weight and narrow M_w/M_n. In 2008, Raynaud et al. [100] and Scholten et al. [101] independently reported that *N*-heterocyclic carbene (NHC) efficiently catalyzed the GTP of methyl methacrylate (MMA) and *tert*-butyl acrylate, as shown in Scheme 1.17a. In addition, Chen and Zang reported that the triphenylmethyl cation oxidatively activated MTS to promote the GTP of MMA, as shown in Scheme 1.17b [102]. Moreover, Kakuchi et al. reported that bis(trifluoromethanesulfonyl) imide (Tf_2NH) [103–107], a very strong Brønsted acid, was capable of promoting the GTP of MMA using MTS as the initiator to afford the narrow dispersed and highly syndiotactic poly(methyl methacrylate) without any side reactions, as shown in Scheme 1.17c [108]. The use of these new catalysts was considered to be the key to realize the GTP of acrylamide monomers proceeding in a living fashion and provide an easy and versatile method to synthesize thermoresponsive polyacrylamides with well-defined structure.

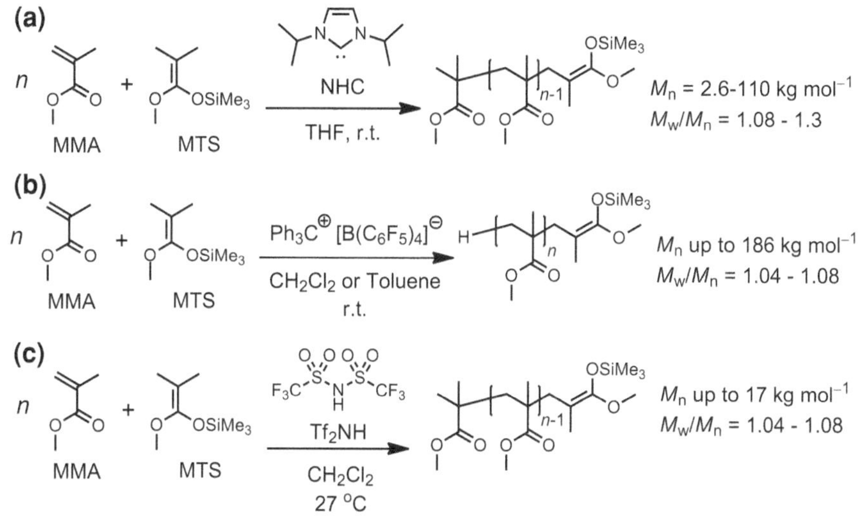

Scheme 1.17 Group transfer polymerization (*GTP*) using a new class of catalysts, such as **a** *N*-heterocyclic carbene (*NHC*), **b** the triphenylmethyl salt, and **c** bis(trifluoromethanesulfonyl) imide (Tf_2NH)

1.6 Objects and Outline of the Thesis

As described in the previous sections, the thermoresponsive polymers have attracted much attention due to their potential applications as intelligent materials. Among the various thermoresponsive polymers, polyacrylamides have been intensively investigated over the past few decades due to their ease of molecular design and synthesis. In order to design intelligent materials with the desired properties, the relationship between the structure and the thermoresponsive properties of the polyacrylamides must be elucidated through developing sophisticated methods to precisely synthesize polyacrylamides with a well-defined primary structure. Thus, this thesis focused on developing easy and versatile methods for the precise synthesis of polyacrylamides with novel primary structures as well as the evaluation of their thermoresponsive properties. The synthesis of azido end-functionalized PNIPAM by the ATRP and the postpolymerization modification using the click reaction was considered to provide a versatile method to synthesize a PNIPAM with a well-defined terminal structure. The group transfer polymerization (GTP) for acrylamide monomers was expected to provide excellent control of the polymerization in comparison to the existing controlled/living polymerizations and become a robust method to precisely synthesize thermoresponsive polyacrylamides and their block copolymers, namely thermoresponsive amphiphilic block copolymers and double-hydrophilic block copolymers.

The outline of this thesis is as follows.

Chapter 2 describes the precise synthesis of urea end-functionalized PNIPAMs and the control of the thermoresponsive properties of PNIPAM by changing the

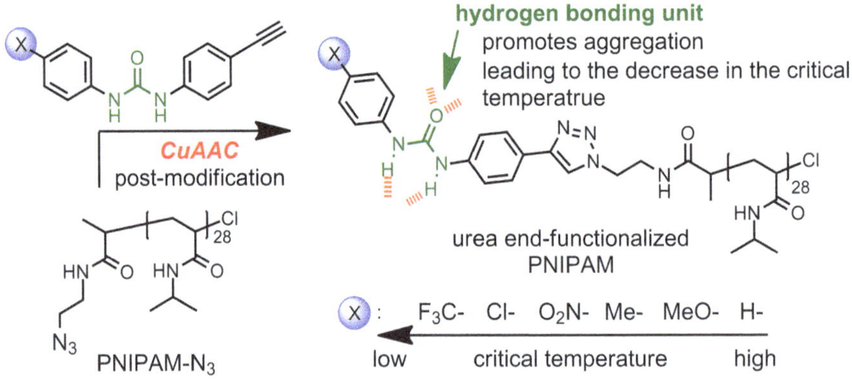

Fig. 1.4 Graphical abstract of Chap. 2

hydrogen bonding ability of the terminal urea groups, as depicted in Fig. 1.4. A series of PNIPAMs with different diphenylurea groups at the chain end (X-Ph-NH-CO-NH-Ph-trz-PNIPAM: X = H, OCH_3, CH_3, NO_2, Cl, and CF_3) were synthesized by the ATRP and the CuAAC. The critical temperature of the obtained polymers varied depending on the hydrogen bonding ability of the introduced urea group. The ^{1}H NMR measurement suggested that the urea end-functionalized PNIPAM was already forming aggregates in water even at a temperature below the critical temperature due to the intermolecular hydrogen bonding of the terminal urea group. The aggregated particles of the resulting polymer were directly characterized by dynamic light scattering and transmission electron microscopy in water at a temperature below its critical temperature. The hydrogen bonding properties of the chain end urea group was concluded to be involved in the aggregation of the PNIPAM in water, leading to the variation in its critical temperature.

Chapter 3 describes the GTP of DMAA promoted by Tf_2NH, one of the strong Brønsted acids, proceeding in a living fashion for the synthesis of polyacrylamides with well-defined structures, as depicted in Fig. 1.5. The combined use of (*Z*)-1-dimethylamino-1-trimethylsiloxy-1-propene [(*Z*)-DATP], an amino silyl enolate, and Tf_2NH realized a highly efficient GTP initiating system for DMAA. The Tf_2NH-promoted GTP of DMAA initiated by (*Z*)-DATP at 0 °C homogeneously proceeded to produce PDMAA with a predetermined molecular weight and a narrow molecular weight distribution. The living nature of the polymerization was confirmed by the matrix-assisted laser desorption/ionization time-of-flight mass spectrometry (MALDI-TOF MS) analysis, kinetic measurements, and a post polymerization experiment.

Chapter 4 describes the development of the easy and versatile method to synthesize thermoresponsive amphiphilic block copolymers and double-hydrophilic block copolymers having a PDEAA segment through the GTP of DEAA, as depicted in Fig. 1.6. The precise synthesis of PDEAA and hydroxyl end-functionalized PDEAA (PDEAA-OH) was achieved by the Tf_2NH-promoted GTP of

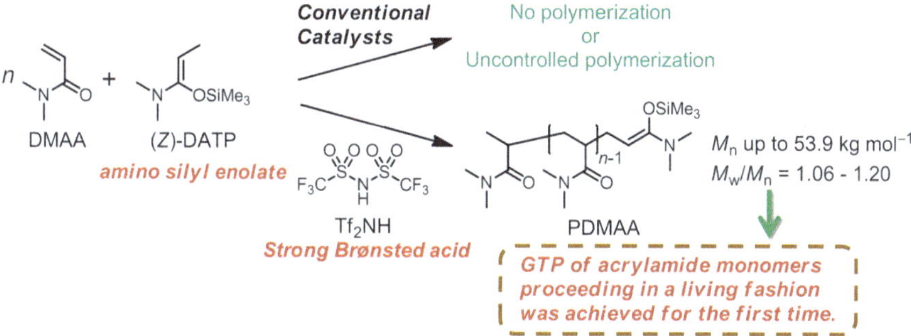

Fig. 1.5 Graphical abstract of Chap. 3

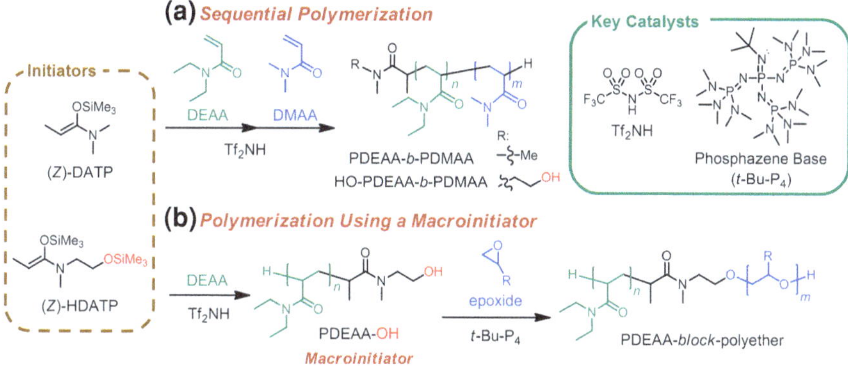

Fig. 1.6 Graphical abstract of Chap. 4

DEAA using an amino silyl enolate, such as (*Z*)-DATP or (*Z*)-1-(*N*-methyl-*N*-(2-trimethylsiloxyethyl)amino)-1-trimethylsiloxy-1-propene [(*Z*)-HDATP], as the initiator in a similar manner to the GTP of DMAA. The synthesis of the double-hydrophilic block copolymer consisting of PDEAA and PDMAA segments was achieved by the sequential polymerization of DEAA and DMAA. The anionic ring-opening polymerization of epoxides, such as ethylene oxide (EO), propylene oxide (PO), 1,2-butene oxide (BO), and 1-ethoxyethyl glycidyl ether (EEGE), using PDEAA-OH as the macroinitiator and a phosphazene base, 1-*tert*-butyl-4,4,4-tris(dimethylamino)-2,2-bis-[tris(dimethylamino)-phosphoranylidenamino]-$2\Lambda^5,4\Lambda^5$-catenadi(phosphazene) (*t*-Bu-P_4), as the catalyst provided a versatile method to synthesize thermoresponsive amphiphilic block polymers and double-hydrophilic block copolymers consisting of a PDEAA segment and a polyether segment. The turbidimetric analysis proved that the thermoresponsive properties of the obtained PDEAA-*b*-polyethers were successfully controlled as originally expected from the molecular design.

References

1. Coughlan DC, Quilty FP, Corrigan OI (2004) Effect of drug physicochemical properties on welling/deswelling kinetics and pulsatile drug release from thermoresponsive poly(*N*-isopropylacrylamide) hydrogels. J Control Rel 98:97–114
2. Tsuda Y, Yamato M, Kikuchi A, Watanabe M, Chen GP, Takahashi Y, Okano T (2007) Thermoresponsive microtextured culture surfaces facilitate fabrication of capillary networks. Adv Mater 19:3633–3636
3. De P, Li M, Gondi SR, Sumerlin BS (2008) Temperature-Regulated activity of responsive polymer−protein conjugates prepared by grafting-from via RAFT polymerization. J Am Chem Soc 130:11288–11289
4. Klouda L, Mikos AG (2008) Thermoresponsive hydrogels in biomedical applications—a review. Eur J Pharm Biopharm 68:34–45
5. Ito T, Hioki T, Yamaguchi T, Shinbo T, Nakao S, Kimura S (2002) Development of a molecular recognition ion gating membrane and estimation of Its pore size control. J Am Chem Soc 124:7840–7846
6. Nykanen A, Nuopponen M, Laukkanen A, Hirvonen SP, Rytela M, Turunen O, Tenhu H, Mezzenga R, Ikkala O, Ruokolainen J (2007) Phase behavior and temperature-responsive molecular filters based on self-assembly of polystyrene-*block*-poly(*N*-isopropylacrylamide)-*block*-polystyrene. Macromolecules 40:5827–5834
7. Tokuyama H, Iwama T (2007) Temperature-swing solid-phase extraction of heavy metals on a poly(*N*-isopropylacrylamide) hydrogel. Langmuir 23:13104–13108
8. Wang H, An YL, Huang N, Ma RJ, Li JB, Shi LQ (2008) Contractive polymeric complex micelles as thermo-sensitive nanopumps. Macromol Rapid Commun 29:1410–1414
9. Schild HG (1992) Poly(*N*-isopropylacrylamide): experiment, theory and application. Prog Polym Sci 17:163–249
10. Idziak I, Avoce D, Lessard D, Gravel D, Zhu XX (1999) Thermosensitivity of aqueous solutions of poly(*N*,*N*-diethylacrylamide). Macromolecules 32:1260–1263
11. Lutz J-F, Akdemir O, Hoth A (2006) Point by point comparison of two thermosensitive polymers exhibiting a similar LCST: Is the age of poly(NIPAM) over? J Am Chem Soc 128:13046–13047
12. Lutz J-F, Andrieu J, Uzgun S, Rudolph C, Agarwal S (2007) Biocompatible, thermoresponsive, and biodegradable: Simple preparation of "all-in-one" biorelevant polymers. Macromolecules 40:8540–8543
13. Lutz J-F, Weichenhan K, Akdemir O, Hoth A (2007) About the phase transitions in aqueous solutions of thermoresponsive copolymers and hydrogels based on 2-(2-methoxyethoxy) ethyl methacrylate and oligo(ethylene glycol) methacrylate. Macromolecules 40:2503–2508
14. Maeda Y (2001) IR spectroscopic study on the hydration and the phase transition of poly(vinyl methyl ether) in water. Langmuir 17:1737–1742
15. Maeda Y, Yamauchi H, Fujisawa M, Sugihara S, Ikeda I, Aoshima S (2007) Infrared spectroscopic investigation of poly(2-methoxyethyl vinyl ether) during thermosensitive phase separation in water. Langmuir 23:6561–6566
16. Fuse C, Okabe S, Sugihara S, Aoshima S, Shibayama M (2004) Water-induced self-assembling of solvent-sensitive block copolymer. Macromolecules 37:7791–7798
17. Okabe S, Sugihara S, Aoshima S, Shibayama M (2003) Heat-induced self-assembling of thermosensitive block copolymer. Rheology and dynamic light scattering study. Macromolecules 36:4099–4106
18. Labbé A, Carlotti S, Deffieux A, Hirao A (2007) Controlled polymerization of glycidyl methyl ether Initiated by onium salt/triisobutylaluminum and investigation of the polymer LCST. Macromol Symp 250:392–397
19. Ogura M, Tokuda H, Imabayashi S-I, Watanabe M (2007) Preparation and solution behavior of a thermoresponsive diblock copolymer of poly(ethyl glycidyl ether) and poly(ethylene oxide). Langmuir 23:9429–9434

20. Inoue S, Kakikawa H, Nakadan N, Imabayashi S-I, Watanabe M (2009) Thermal response of poly(ethoxyethyl glycidyl ether) grafted on gold surfaces probed on the basis of temperature-dependent water wettability. Langmuir 25:2837–2841
21. Hoogenboom R (2009) Poly(2-oxazoline)s: A polymer class with numerous potential applications. Angew Chem Int Ed 48:7978–7994
22. Lau ACW, Wu C (1999) Thermally sensitive and biocompatible poly(*N*-vinylcaprolactam): Synthesis and characterization of high molar mass linear chains. Macromolecules 32:581–584
23. Laukkanen A, Valtola L, Winnik FM, Tenhu H (2004) Thermally sensitive and biocompatible poly(*N*-vinylcaprolactam): Synthesis and characterization of high molar mass linear chains. Macromolecules 37:2268–2274
24. Ieong NS, Redhead M, Bosquillon C, Alexander C, Kelland M, O'Reilly RK (2011) The missing lactam-thermoresponsive and biocompatible poly(*N*-vinylpiperidone) polymers by xanthate-mediated RAFT polymerization. Macromolecules 44:886–893
25. Arotçaréna M, Heise B, Ishaya S, Laschewsky A (2002) Switching the inside and the outside of aggregates of water-soluble block copolymers with double thermoresponsivity. J Am Chem Soc 124:3787–3793
26. Soo PL, Eisenberg A (2004) Preparation of block copolymer vesicles in solution. J Polym Sci, Part B: Polym Phys 42:923–938
27. Rodríguez-Hernández J, Chécot F, Gnanou Y, Lecommandoux S (2005) Toward 'smart' nano-objects by self-assembly of block copolymers in solution. Prog Polym Sci 30:691–724
28. Lutz J-F (2006) Solution self-assembly of tailor-made macromolecular building blocks prepared by controlled radical polymerization techniques. Polym Int 55:979–993
29. de las Heras Alarcón C, Pennadam S, Alexander C (2005) Stimuli responsive polymers for biomedical applications. Chem Soc Rev 34:276-285
30. Cölfen H (2001) Double-hydrophilic block copolymers: Synthesis and application as novel surfactants and crystal growth modifiers. Macromol Rapid Commun 22:219–252
31. Gil ES, Hudson SM (2004) Stimuli-reponsive polymers and their bioconjugates. Prog Polym Sci 29:1173–1222
32. Riess G (2003) Micellization of block copolymers. Prog Polym Sci 28:1107–1170
33. Sundararaman A, Stephan T, Grubbs RB (2008) Reversible restructuring of aqueous block copolymer assemblies through stimulus-induced changes in amphiphilicity. J Am Chem Soc 130:12264–12266
34. Cao Y, Zhao N, Wu K, Zhu XX (2009) Solution properties of a thermosensitive triblock copolymer of *N*-alkyl substituted acrylamides. Langmuir 25:1699–1704
35. Xia Y, Yin X-C, Bruke NAD, Stöver HDH (2005) Thermal response of narrow-disperse poly(*N*-isopropylacrylamide) prepared by atom transfer radical polymerization. Macromolecules 38:5937–5943
36. Kujawa P, Segui F, Shaban S, Diab C, Okada Y, Tanaka F, Winnik FM (2006) Impact of end-group association and main-chain hydration on the thermosensitive properties of hydrophobically modified telechelic poly(*N*-isopropylacrylamides) in water. Macromolecules 39:341–348
37. Xie X-Y, Hogen-Esch TE (1996) Anionic synthesis of narrow molecular weight distribution water-soluble poly(*N*,*N*-dimethylacrylamide) and poly(*N*-acryloyl-*N*′-methylpiperazine). Macromolecules 29:1746–1752
38. Kobayashi M, Okuyama S, Ishizone T, Nakahama S (1999) Stereospecific anionic polymerization of *N*,*N*-dialkylacrylamides. Macromoelcules 32:6466–6477
39. Kobayashi M, Ishizone T, Nakahama S (2000) Additive effect of triethylborane on anionic polymerization of *N*,*N*-dimethylacrylamide and *N*,*N*-diethylacrylamide. Macomolecules 33:4411–4416
40. Ishizone T, Ito M (2002) Synthesis of well-defined poly(*N*-isopropylacrylamide) by the anionic polymerization of *N*-methoxymethyl-*N*-isopropylacrylamide. J Polym Sci, Part A: Polym Chem 40:4328–4332

41. Ito M, Ishizone T (2006) Living anionic polymerization of *N*-methoxymethyl-*N*-isopropylacrylamide: Synthesis of well-defined poly(*N*-isopropylacrylamide) having various stereoregularity. J Polym Sci, Part A: Polym Chem 44:4832–4845
42. Mariott WR, Chen EY-X (2004) Stereospecific, coordination polymerization of acrylamides by chiral *ansa*-metallocenium alkyl and ester enolate cations. Macromolecules 37:4741–4743
43. Mariott WR, Chen EY-X (2005) Mechanism and scope of stereospecific, coordinative-anionic polymerization of acrylamides by chiral zirconocenium ester and amide enolates. Macromolecules 38:6822–6832
44. Miyake GM, Chen EY-X (2008) Metallocene-mediated asymmetric coordination polymerization of polar vinyl monomers to optically active, stereoregular polymers. Macromolecules 41:3405–3416
45. Habaue S, Isobe Y, Okamoto Y (2002) Stereocontrolled radical polymerization of acrylamides and methacrylamides using Lewis acids. Tetrahedron 58:8205–8209
46. Okamoto Y, Habaue S, Isobe Y, Suito Y (2003) Stereocontrol using lewis acids in radical polymerization. Macromol Symp 195:75–80
47. Ray B, Isobe Y, Morioka K, Habaue S, Okamoto Y, Kamigaito M, Sawamoto M (2003) Synthesis of isotactic poly(*N*-isopropylacrylamide) by RAFT polymerization in the presence of Lewis acid. Macromolecules 36:543–545
48. Ray B, Isobe Y, Matsumoto K, Habaue S, Okamoto Y, Kamigaito M, Sawamoto M (2004) RAFT polymerization of *N*-isopropylacrylamide in the absence and presence of $Y(OTf)_3$: Simultaneous control of molecular weight and tacticity. Macromolecules 37:1702–1710
49. Xia Y, Bruke AD, Stöver HDH (2006) End group effect on the thermal response of narrow-disperse poly(*N*-isopropylacrylamide) prepared by Atom Transfer Radical Polymerization. Macromolecules 39:2275–2283
50. Duan Q, Miura Y, Narumi A, Shen X-D, Sato S-I, Satoh T, Kakuchi T (2006) Synthesis and thermoresponsive property of end-functionalized poly(*N*-isopropylacrylamide) with pyrenyl group. J Polym Sci, Part A: Polym Chem 44:1117–1124
51. Duan Q, Narumi A, Miura Y, Shen X-D, Sato S-I, Satoh T, Kakuchi T (2006) Thermoresponsive property controlled by end-functionalization of poly(*N*-isopropylacrylamide) with phenyl, biphenyl, and triphenyl groups. Polym J 38:306–310
52. Plummer R, Hill DJT, Whittaker AK (2006) Solution properties of star and linear poly(*N*-isopropylacrylamide). Macromolecules 39:8379–8388
53. Vogt AP, Sumerlin BS (2008) Tuning the temperature response of branched poly(*N*-isopropylacrylamide) prepared by RAFT Polymerization. Macromolecules 41:7368–7373
54. Qiu X-P, Tanaka F, Winnik FM (2007) The temperature-induced phase separation of well-defined cyclic poly(*N*-isopropylacrylamides) in aqueous solution. Macromolecules 40:7069–7071
55. Xu J, Ye J, Liu S (2007) Synthesis of well-defined cyclic poly(*N*-isopropylacrylamide) via click chemistry and Its unique thermal phase transition behavior. Macromolecules 40:9103–9110
56. Ye J, Xu J, Hu J, Wang X, Zhang G, Liu S, Wu C (2008) Comparative study of temperature-induced association of cyclic and linear poly(*N*-isopropylacrylamide) chains in dilute solutions by laser light scattering and stopped-flow temperature jump. Macromolecules 41:4416–4422
57. Satokawa Y, Shikata T, Tanaka F, Qiu X-P, Winnik FM (2009) Hydration and dynamic behavior of a cyclic poly(*N*-isopropylacrylamide) in aqueous solution: Effects of the polymer chain topology. Macromolecules 42:1400–1403
58. Hawker CJ, Bosman AW, Harth E (2001) New polymer synthesis by nitroxide mediated living radical polymerizations. Chem Rev 101:3661–3688
59. Matyjaszewski K, Xia J-H (2001) Atom transfer radical polymerization. Chem Rev 101:2921–2990

60. Ouchi M, Terashima T, Sawamoto M (2009) Transition metal-catalyzed living radical polymerization: Toward perfection in catalysis and precision polymer synthesis. Chem Rev 109:4963–5050
61. Moad G, Rizzard E, Thang SH (2009) Living radical polymerization by the RAFT process – A second update. Aust J Chem 62:1402–1472
62. Semsarilar M, Perrier S (2010) 'Green' Reversible Addition-Fragmentation Chain-Transfer (RAFT) polymerization. Nature Chem 2:811–820
63. Braunecker WA, Matyjaszewski K (2007) Controlled/living radical polymerization: Features, developments, and perspectives. Prog Polym Sci 32:93–146
64. Benoit D, Chaplinski V, Braslau R, Hawker CJ (1999) Development of a universal alkoxyamine for "living" free radical polymerizations. J Am Chem Soc 121:3904–3920
65. Binder WH, Gloger D, Weinstabl H, Allmaier G, Pittenauer E (2007) Telechelic poly(*N*-isopropylacrylamides) via nitroxide-mediated controlled polymerization and "click" chemistry: Livingness and "grafting-from" methodology. Macromolecules 40:3097–3107
66. Teodorescu M, Matyjaszewski K (2000) Functionalization of polymers prepared by ATRP using radical addition reactions. Macromol Rapid Commun 21:190–194
67. Neugebauer D, Matyjaszewski K (2003) Copolymerization of *N*,*N*-dimethylacrylamide with n-butyl acrylate via atom transfer radical polymerization. Macromolecules 36:2598–2603
68. Masci G, Giacomelli L, Cresenzi V (2004) Atom transfer radical polymerization of N-isopropylacrylamide. Macromol Rapid Commun 25:559–564
69. Schilli C, Lanzendörfer MG, Müller AHE (2002) Benzyl and cumyl dithiocarbamates as chain transfer agent in the RAFT polymerization of *N*-isopropylacrylamide. In situ FT-NIR and MALDI-TOF MS investigation. Macromolecules 35:6819–6827
70. Convertine AJ, Ayres N, Scales CW, Lowe AB, McCormick CL (2004) Facile, controlled, room-temperature RAFT polymerization of *N*-isopropylacrylamide. Biomacromolecules 5:1177–1180
71. Yusa S-I, Shimada Y, Mitsukami Y, Yamamoto T, Morishima Y (2004) Heat-induced association and dissociation behavior of amphiphilic diblock copolymers synthesized via reversible addition-fragmentation chain transfer radical polymerization. Macromolecules 27:7505–7513
72. Convertine AJ, Lokitz BS, Vasileva Y, Myrick L, Scales CW, Lowe AB, McCormick CL (2006) Direct synthesis of thermally responsive DMA/NIPAM diblock and DMA/NIPAM/DMA triblock copolymers via aqueous, room temperature RAFT polymerization. Macromolecules 39:1724–1730
73. Ganachaud F, Monteiro M, Gilbert RG, Dourges M, Thang SH, Rizzardo E (2000) Molecular weight characterization of poly(*N*-isopropylacrylamide) prepared by living free-radical polymerization. Macromolecules 33:6738–6745
74. Kolb HC, Finn MG, Sharpless KB (2001) Click chemistry: Diverse chemical function from a few good reactions. Angew Chem Int Ed 40:2004–2021
75. Lutz J-F (2007) 1,3-dipolar cycloadditions of azides and alkynes: A universal ligation tool in polymer and materials science. Angew Chem Int Ed 46:1018–1025
76. Binder WH, Sachsenhofer R (2008) 'Click' chemistry in polymer and material science: An update. Macromol Rapid Commun 29:952–981
77. Mansfeld U, Pietsch C, Hoogenboom R, Becer R, Schubert US (2010) Clickable initiators, monomers and polymers in controlled radical polymerizations – a prospective combination in polymer science. Polym Chem 1:1560–1598
78. Hoyle CE, Lowe AB, Bowmann CN (2010) Thiol-click chemistry: a multifaceted toolbox for small molecule and polymer synthesis. Chem Soc Rev 39:1355–1387
79. Xu J, Liu S (2008) Synthesis of Well-Defined 7-Arm and 21-Arm Poly(*N*-isopropylacrylamide) Star Polymers with β-Cyclodextrin Cores via Click Chemistry and Their Thermal Phase Transition Behavior in Aqueous Solution. J Polym Sci, Part A: Polym Chem 47:404–419

80. André X, Zhang M, Müller AHE (2005) Thermo- and pH-responsive micelles of poly(acrylic acid)-block-poly(*N*,*N*-diethylacrylamide). Macromol Rapid Commun 26:558–563
81. Vinogradova L, Fedorova L, Adler H-JP, Kuckling D, Seifert D, Tsvetanov CB (2005) Controlled anionic block copolymerization with *N*,*N*-dialkylacrylamide as a second block. Macromol Chem Phys 206:1126–1133
82. Xu Y, Shi L, Ma R, Zhang W, An Y, Zhu XX (2007) Synthesis and micellization of thermo- and pH-responsive block copolymer of poly(*N*-isopropylacrylamide)-*block*-poly(4-vinylpyridine). Polymer 48:1711–1717
83. Klaikherd A, Nagamani C, Thayumanavan S (2009) Multi-stimuli sensitive amphiphilic block copolymer assemblies. J Am Chem Soc 131:4830–4838
84. Webster OW, Hertler WR, Sogah DY, Farnham WB, RajanBabu TV (1983) Group-transfer polymerization. 1. A new concept for addition polymerization with organosilicon initiators. J Am Chem Soc 105:5706–5708
85. Sogah DY, Hertler WR, Webster OW, Cohen GM (1987) Group transfer polymerization—polymerization of acrylic monomers. Macromolecules 20:1473–1488
86. Webster OW (2004) Group transfer polymerization: Mechanism and comparison with other methods for controlled polymerization of acrylic monomers. Adv Polym Sci 167:1–34
87. Schubert W, Bandermann F (1989) Group trabsfer polymerization of methyl acrylate in acetonitrile. Makromol Chem 190:2161–2171
88. Schubert W, Sitz HD, Bandermann F (1989) Group transfer polymerization of methyl-methacrylate and methyl acrylate in tetrahydrofuran with tris(piperidino)sulfonium bifluoride as catalyst. Makromol Chem 190:2193–2201
89. Schubert W, Bandermann F (1989) Group Transfer polymerization of methyl acrylate in tetrahydrofuran with tetrabutylammonium fluoride trihydrate and tetrabutylammonium cyanide as catalysts. Makromol Chem 190:2721–2726
90. Dicker IB, Cohen GM, Farnham WB, Hertler WR, Laganis ED, Sogah DY (1990) Catalyzed silicon-mediated living polymerizations—structure control, catalyst design and mechanisms. Macromolecules 23:4034–4041
91. Patrickios CS, Hertler WR, Abbott NL, Hatton TA (1994) Diblock, ABC triblock, and random methacryloc polyampholytes—synthesis by Group-transfer polymerization and solution behavior. Macromolecules 27:930–937
92. Hertler WR, Sogah DY, Webster OW (1984) Group-transfer polymerization. 3. lewis acid catalysis. Macromolecules 17:1415–1417
93. Dicker IB (1988) Group transfer polymerization (GTP) of acrylates catalyzed by mercuric iodide. Polym Prep (Am Chem Soc, Div Polym Chem) 29-2:114–115
94. Zhuang R, Müller AHE (1994) Kinetics and mechanism of Group-transfer polymerization of *n*-butyl acrylate catalyzed by $HgI_2/(CH_3)_3SiI$ in toluene. Macromol Symp 85:379–392
95. Zhuang R, Müller AHE (1995) Group-transfer polymerization of *n*-butyl acrylate with lewis-acid catalysis. 1.Kinetic investigation using HgI_2 as a catalyst in toluene. Macromolecules 28:8035–8042
96. Zhuang R, Müller AHE (1995) Group-transfer polymerization of *n*-butyl acrylate with lewis-acid catalysis. 2. Kinetic investigation using the HgI_2/Me_3SiI catalyst system in toluene and methylene-chloride. Macromolecules 28:8043–8050
97. Eggert M, Freitag R (1994) Poly-*N*,*N*-diethylacrylamide prepared by group transfer polymerization: Synthesis, characterization, and solution properties.J Polym Sci, Part A: Polym Chem 32:803–813
98. Freitag R, Baltes T, Eggert M (1994) A comparison of thermoreactive water-soluble poly-*N*,*N*-diethylacrylamide prepared by anionic and by group transfer polymerization. J Polym Sci, Part A: Polym Chem 32:3019–3030
99. Baltes T, Garret-Flaudy F, Freitag R (1999) Investigation of the LCST of polyacrylamides as a function of molecular parameters and the solvent composition. J Polym Sci, Part A: Polym Chem 37:2977–2989

100. Raynaud J, Ciolino A, Baceiredo A, Destarac M, Bonnette F, Kato T, Gnanou Y, Taton D (2008) Harnessing the potential of *N*-heterocyclic carbenes for the rejuvenation of Group-transfer polymerization of (Meth)acrylics. Angew Chem Int Ed 47:5390–5393
101. Scholten MD, Hedrick JL, Waymouth RM (2008) Group transfer polymerization of acrylates catalyzed by *N*-heterocyclic carbenes. Macromolecules 41:7399–7404
102. Zhang Y-T, Chen EY-X (2008) Controlled polymerization of methacrylates to high molecular weight polymers using oxidatively activated group transfer polymerization initiators. Macromolecules 41:36–42
103. Foropoulos J, Desmarteau DD (1984) Synthesis, properties, and reactions of bis((trifluoromethyl)sulfonyl)imide, $(CF_3SO_2)_2NH$. Inorg Chem 23:3720–3723
104. Ishihara K, Hiraiwa Y, Yamamoto H (2001) A high yield procedure for the Me_3SiNTf_2-induced carbon-carbon bond-forming reactions of silyl nucleophiles with carbonyl compounds: The importance of addition order and solvent effects. Synlett 1851–1854
105. Hiraiwa Y, Ishihara K, Yamamoto H (2006) Crucial role of the conjugate base for silyl lewis acid induced mukaiyama aldol reactions. Eur J Org Chem 2006:1837–1844
106. Akiyama T (2007) Stronger brønsted acids. Chem Rev 107:5744–5758
107. Yamamoto H (2007) New reaction and new catalyst—a personal perspective. Tetrahedron 63:8377–8412
108. Kakuchi R, Chiba K, Fuchise K, Sakai R, Satoh T, Kakuchi T (2009) Strong brønsted acid as a highly efficient promoter for group transfer polymerization of methyl methacrylate. Macromolecules 42:8747–8750

Chapter 2
Control of Thermoresponsive Properties of Urea End-Functionalized Poly(*N*-isopropylacrylamide) Based on the Hydrogen Bond Assisted Self-Assembly in Water

2.1 Introduction

The hydrogen bond has been recognized as one of the representative non-covalent interactions for realizing a three-dimensionally organized molecular architecture. In fact, biological macromolecular systems ingeniously construct specific higher-order structures that are stable in an aqueous medium, in which multiple hydrogen bonds are indispensable for stabilizing the structures. Although there are a number of artificial molecular architectures fabricated through hydrogen bonding in organic solvents, the realization of a supramolecular assembly based on hydrogen bonding in water has still remained an interesting issue because the intermolecular hydrogen bonds are easily disrupted in protic and polar solvents, such as water.

For the supramolecular assembly based on hydrogen bonding in water, the hydrophobic microenvironment approach that is inspired by a biological system have been developed for protecting the hydrogen bonding capability from disruptive solvation by water molecules, thus making it possible to construct a well-ordered self-assembly from synthetic molecules in water [1–3]. For example, Meijer et al. revealed that bifunctional ureidotriazines can self-assemble to form a helical supramolecular polymer in water through cooperative hydrogen bonds that are shielded by the hydrophobic microenvironment [4]. Kimizuka et al. achieved fabrication of a hierarchically self-assembled bilayer membrane in water through complementary hydrogen-bond pairs [5]. Sijbesma et al. further succeeded in the formation of the self-assembly from an amphiphilic triblock copolymer in water, in which urea groups that are placed at the center of a nonpolar segment contribute to the strong intermolecular hydrogen bonding [6]. Despite the definite validity of these approaches, the construction of stable intermolecular hydrogen bonding in water has still been a challenging task because an accurate and strict molecular design is required for the functional groups involved in the intermolecular hydrogen bonding, hydrophilic chains for water-solubility, and a hydrophobic microenvironment for realization of the hydrogen bonding properties. Additionally, only the construction of the self-assembly with a specific structure and morphology

K. Fuchise, *Design and Precise Synthesis of Thermoresponsive Polyacrylamides*, Springer Theses, DOI: 10.1007/978-4-431-55046-4_2,

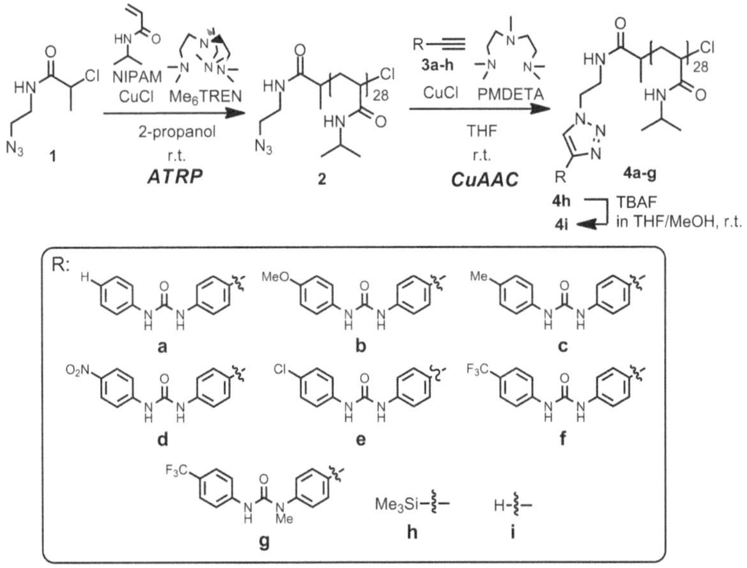

Scheme 2.1 Synthesis of various urea end-functionalized PNIPAMs (**4a–g**) by a combination of the ATRP and the copper(I) catalyzed azide-alkyne cycloaddition (CuAAC)

has been achieved by the hydrogen bonding in water. Therefore, the control of the characteristic feature and function of the self-assembled polymer by adjusting its hydrogen bonding ability in water will be significant.

As one of the demonstrations of the macromolecular assembly based on the hydrogen bonding in water, the author now focuses on controlling the thermoresponsive properties of the urea end-functionalized poly(*N*-isopropylacrylamide) (PNIPAM) by adjusting the hydrogen bond formation between the chain end urea groups. PNIPAM is a representative thermoresponsive polymer and exhibits a coil to globule transition of the polymer structure in water at a temperature known as the lower critical solution temperature (LCST), which is generally around 32 °C [7]. Due to its versatility, many applications are expected in the diverse fields of material science including bioengineering [8–11] and nanotechnology [12–15]. Currently, it has been clarified that the cloud point of PNIPAM varies depending on its primary structure, such as molecular weight [16–18], stereoregularity [19–22], and chain end structure [17, 18, 23–26]. However, there is no report about the influence of the intermolecular hydrogen bonding of the end-functional group in PNIPAM on its cloud point.

In this chapter, the author describes the control of the cloud point of PNIPAM using the hydrogen bonding of urea groups in the polymer chain end. The well-defined PNIPAMs with urea groups as the end-functionality were prepared using a combination of the atom transfer radical polymerization (ATRP) and the copper(I) catalyzed azide-alkyne cycloaddition (CuAAC), as illustrated in Scheme 2.1. First, the azido end-functionalized PNIPAM (**2**) was synthesized by the ATRP of NIPAM using *N*-(2-azidoethyl)-2-chloropropionamide (**1**). The CuAAC between **2** and the diphenylurea derivatives (**3a–f**) was subsequently conducted for conjugating the various urea groups, producing PNIPAMs bearing a series of chain end urea groups (**4a–f**). The urea groups to be introduced at the chain end of the PNIPAM were expected to impart a hydrogen bonding ability to the polymer, leading to its hierarchical self-assembly [6, 27]. Additionally, the diphenylurea derivatives were designed for facilitating the efficient formation of intermolecular hydrogen bonds between the urea groups in water. The relationship between the changes in the cloud point of the resulting polymers and the hydrogen bonding ability of the various urea groups was clarified by a turbidimetric analysis. In addition, in order to clarify the influence of the hydrogen bond on the cloud point, PNIPAMs bearing *N*-methylated urea (**4g**) and triazolyl groups (**4i**) were also synthesized, as shown in Schemes 2.1, and their cloud points were compared to those of **4a–f**. Dynamic light scattering (DLS) and transmission electron microscopy (TEM) measurements were performed for the direct observation of the polymer assembly in water, providing an insight into the variation of the cloud point based on the supramolecular assembly.

2.2 Experimental Section

2.2.1 Materials

N-Isopropylacrylamide (NIPAM) was kindly supplied from the Kohjin Co., Japan, and recrystallized twice from hexane/toluene (10/1, v/v) prior to use. *N*-(2-Azidoethyl)-2-chloropropionamide (**1**) was prepared according to our previous report [18]. Tris[2-(dimethylamino)ethyl]amine (Me_6TREN) was donated by the Mitsubishi Chemical Co., Japan, and distilled over CaH_2 under reduced pressure. 2-Propanol (HPLC Grade), tetrahydrofuran (THF), phenyl isocyanate (98 %), dichloromethane, ethyl acetate, and acetone were purchased from Kanto Chemicals Co., Inc., and used as received. Acetonitrile was purchased from Kanto Chemicals, and dried over activated molecular sieves 4A for one day, and then distilled over CaH_2. 4-Ethynylaniline was purchased from Wako Pure Chemical Industries, Ltd., and used without further purification. *N*-Methyl-4-ethynylaniline was prepared according to the literature [28]. 4-(Trifluoromethyl) phenyl isocyanate (98 %), 4-methoxyphenyl isocyanate (98 %), 4-nitrophenyl isocyanate (98 %), trimethylsilylacetylene (98 %), and *N,N,N′,N″,N″*-pentamethyldiethylenetriamine (PMDETA, 98 %) were purchased from Tokyo Kasei Kogyo Co., Ltd., and used as received. 4-Methylphenyl isocyanate (98 %), 4-chlorophenyl isocyanate (98 %), tetra-*n*-butylammonium fluoride (TBAF) in THF (1.0 mol L^{-1}), and copper(I) chloride (CuCl, 99.999 %) were obtained from Aldrich Chemicals Co., Inc., and used as received.

2.2.2 Instruments

The ^{1}H (400 MHz) and ^{13}C NMR (100 MHz) spectra were recorded using a JEOL JNM-A400II instrument. The IR spectra were recorded using a Perkin-Elmer Paragon 1000 FT-IR instrument. Size exclusion chromatography (SEC) was performed at 40 °C using a Jasco high performance liquid chromatography (HPLC) system (PU-980 Intelligent HPLC pump, CO-965 column oven, RI-930 Intelligent RI detector, and Shodex DEGAS KT-16) equipped with a Shodex Asahipak GF-310 HQ column (linear, 7.6 mm × 300 mm; pore size, 20 nm; bead size, 5 μm; exclusion limit, 4 × 10^4) and a Shodex Asahipak GF-7M HQ column (linear, 7.6 mm × 300 mm; pore size, 20 nm; bead size, 9 μm; exclusion limit, 4 × 10^7) in *N*,*N*-dimethylformamide (DMF) containing lithium chloride (0.01 mol L^{-1}) at the flow rate of 0.4 mL min^{-1}. The number-average molecular weight ($M_{n,SEC}$) and polydispersity index (M_w/M_n) of the synthesized polymers were determined on the basis of a polystyrene calibration. The ultraviolet-visible (UV-vis) spectra were measured using a 10-mm path length cell by a Jasco V-550 spectrophotometer equipped with an EYELA NCB-1200 temperature controller and a Jasco ETC-505T temperature controller. Dynamic light scattering (DLS) was measured by an Otsuka Electronics FDLS-3000 light scattering spectrometer with an Otsuka Electronics NM-454L temperature controller. Transmission electron microscopy (TEM) was performed on a JEOL JEM-1230.

2.2.3 Synthesis of 1-(4-Ethynylphenyl)-3-phenylurea (3a)

Phenyl isocyanate (2.30 mL, 21.2 mmol) was added to a solution of 4-ethynylaniline (2.61 g, 22.3 mmol) in dry acetonitrile (47.5 mL) at room temperature under an N_2 atmosphere. The reaction mixture was refluxed for 5 h. After removal of the solvent, the residue was purified by column chromatography on silica gel with acetone/dichloromethane (1/14, v/v) to give **3a** as a solid. Yield = 4.48 g (90 %). ^{1}H NMR (400 MHz, DMSO-d_6, δ): 4.04 (s, 1H, *H*C ≡ C–), 6.98 (t, *J* = 7.3 Hz, 1H, aromatic), 7.29 (t, *J* = 7.3 Hz, 2H, aromatic), 7.39 (d, *J* = 8.8 Hz, 2H, aromatic), 7.45 (d, *J* = 8.4 Hz, 2H, aromatic), 7.48 (d, *J* = 8.8 Hz, 2H, aromatic), 8.72 (s, 1H, –N*H*–CO–),8.87 (s, 1H, –N*H*–CO–). ^{13}C NMR (100 MHz, DMSO-d_6, δ): 79.34, 83.74, 114.53, 117.87, 118.30, 122.01, 128.78, 132.43, 139.42, 140.38, 152.24. Anal. Calcd for $C_{15}H_{12}N_2O$ (236.27): C, 76.25; H, 5.12; N, 11.86. Found: C, 76.19; H, 5.12; N, 11.91.

2.2.4 Synthesis of 1-(4-Ethynylphenyl)-3-(4-methoxyphenyl) urea (3b)

A solution of 4-ethynylaniline (1.85 g, 15.8 mmol) in dry THF (15 mL) was added to a solution of 4-methoxyphenyl isocyanate (1.94 mL, 15.1 mmol) in dry THF (35 mL) at room temperature under an N_2 atmosphere. The reaction mixture was stirred for 18 h. After removal of the solvent, **3b** was obtained as a solid. Yield = 2.79 g (69 %). ^{1}H NMR (400 MHz, DMSO-d_6, δ): 3.72 (s, 3H, $-OCH_3$), 4.02 (s, 1H, $HC\equiv C-$), 6.86 (d, 2H, J = 8.9 Hz, aromatic), 7.35 (d, J = 8.9 Hz, 2H, aromatic), 7.38 (d, J = 8.4 Hz, 2H, aromatic), 7.47 (d, J = 8.4 Hz, 2H, aromatic), 8.52 (s, 1H, –N*H*–CO–), 8.79 (s, 1H, –N*H*–CO–). ^{13}C NMR (100 MHz, DMSO-d_6, δ): 55.13, 79.21, 83.77, 113.97, 114.29, 117.74, 120.16, 132.38, 140.57, 152.41, 154.49. Anal. Calcd for $C_{16}H_{14}N_2O_2$ (266.29): C, 72.16; H, 5.30; N, 10.52. Found: C, 71.90; H, 5.36; N, 10.50.

2.2.5 Synthesis of 1-(4-Ethynylphenyl)-3-(4-methylphenyl) urea (3c)

A solution of 4-ethynylaniline (1.97 g, 16.8 mmol) in dry THF (23 mL) was added to a solution of 4-methylphenyl isocyanate (2.01 mL, 16.0 mmol) in dry THF (30 mL) at room temperature under an N_2 atmosphere. The reaction mixture was stirred for 18 h. After removal of the solvent, **3c** was obtained as a solid. Yield = 3.64 g (91 %). ^{1}H NMR (400 MHz, DMSO-d_6, δ): 2.24 (s, 3H, $-CH_3$), 4.02 (s, 1H, $HC\equiv C-$), 7.08 (d, J = 8.4 Hz, 2H, aromatic), 7.33 (d, J = 8.4 Hz, 2H, aromatic), 7.38 (d, J = 8.9 Hz, 2H, aromatic), 7.46 (d, J = 8.9 Hz, 2H, aromatic), 8.61 (s, 1H, –N*H*–CO–), 8.82 (s, 1H, –N*H*–CO–). ^{13}C NMR (100 MHz, DMSO-d_6, δ): 20.29, 79.25, 83.74, 114.39, 117.79, 118.39, 129.14, 130.85, 132.39, 136.82, 140.46, 152.27. Anal. Calcd for $C_{16}H_{14}N_2O$ (250.29): C, 76.78; H, 5.64; N, 11.19. Found: C, 76.83; H, 5.78; N, 11.16.

2.2.6 Synthesis of 1-(4-Ethynylphenyl)-3-(4-nitrophenyl) urea (3d)

A solution of 4-ethynylaniline (2.30 g, 19.6 mmol) in dry THF (15 mL) was added to a solution of 4-nitrophenyl isocyanate (2.92 g, 17.8 mmol) in dry THF (35 mL) at room temperature under an N_2 atmosphere. The reaction mixture was stirred for 43 h. After removal of the solvent, dichloromethane was added to the residue. The soluble part was filtered and washed with 1 mol L^{-1} HCl. The organic layer was dried over anhydrous Na_2SO_4 and evaporated to dryness. The residue was purified by recrystallization from acetonitrile to give **3d** as a yellow solid. Yield = 2.67 g (54 %). ^{1}H NMR (400 MHz, DMSO-d_6, δ): 4.07 (s, 1H, $HC\equiv C-$), 7.42 (d, J = 8.8 Hz, 2H, aromatic), 7.50 (d, J = 8.8 Hz, 2H, aromatic), 7.69 (d, J = 9.3 Hz, 2H, aromatic), 8.20 (d, J = 9.3 Hz, 2H, aromatic), 9.12 (s, 1H, –N*H*–CO–), 9.48 (s, 1H, –N*H*–CO–). ^{13}C NMR (100 MHz, DMSO-d_6, δ): 79.65, 83.58, 115.30, 117.51, 118.34, 125.13, 132.47, 139.67, 141.15, 146.12, 151.75. Anal. Calcd for $C_{15}H_{11}N_3O_3$ (281.27): C, 64.05; H, 3.94; N, 14.94. Found: C, 63.76; H, 4.01; N, 14.93.

2.2.7 *Synthesis of 1-(4-Ethynylphenyl)-3-(4-chlorophenyl) urea (3e)*

A solution of 4-ethynylaniline (1.82 g, 15.5 mmol) in dry THF (20 mL) was added to a solution of 4-chlorophenyl isocyanate (2.27 g, 14.8 mmol) in dry THF (35 mL) at room temperature under an N_2 atmosphere. The reaction mixture was stirred overnight. After removal of the solvent, the residue was stirred in 1 mol L^{-1} HCl, and then filtered. The filtrate was dried in vacuo to give **3e** as a solid. Yield = 3.52 g (87 %). ^{1}H NMR (400 MHz, DMSO-d_6, δ): 4.02 (s, 1H, $HC \equiv C-$), 7.33 (d, J = 8.9 Hz, 2H, aromatic), 7.40 (d, J = 8.8 Hz, 2H, aromatic), 7.48 (d, J = 8.8 Hz, 2H, aromatic), 7.49 (d, J = 9.1 Hz, 2H, aromatic), 8.88 (s, 1H, –N*H*–CO–), 8.93 (s, 1H, –N*H*–CO–). ^{13}C NMR (100 MHz, DMSO-d_6, δ): 79.32, 83.71, 114.73, 118.00, 119.83, 125.58, 128.60, 132.43, 138.45, 140.21, 152.17. Anal. Calcd for $C_{15}H_{11}N_2OCl$ (270.71): C, 66.55; H, 4.10; N, 10.35; Cl, 13.10. Found: C, 66.54; H, 4.13; N, 10.28; Cl, 13.13.

2.2.8 *Synthesis of 1-(4-Ethynylphenyl)-3-[4-(trifluoromethyl) phenyl]urea (3f)*

A solution of 4-ethynylaniline (1.55 g, 17.2 mmol, 1.05 eq) in dry THF (20 mL) was added to a solution of 4-(trifluoromethyl)phenyl isocyanate (2.45 mL, 16.4 mmol) in dry THF (35 mL) at room temperature under an N_2 atmosphere. The reaction mixture was stirred for 14 h. After removal of the solvent, the residue was dissolved in ethyl acetate and washed with 1 mol L^{-1} NaOH, followed by water. The organic layer was dried over anhydrous $MgSO_4$, and evaporated to dryness. The residue was purified by recrystallization from acetonitrile to give **3f** as a solid. Yield = 3.26 g (65 %). ^{1}H NMR (400 MHz, DMSO-d_6, δ): 4.06 (s, 1H, $HC \equiv C-$), 7.41 (d, J = 8.8 Hz, 2H, aromatic), 7.49 (d, J = 8.8 Hz, 2H, aromatic), 7.64 (d, J = 9.4 Hz, 2H, aromatic), 7.67 (d, J = 9.0 Hz, 2H, aromatic), 9.01 (s, 1H, –N*H*–CO–), 9.16 (s, 1H, –N*H*–CO–). ^{13}C NMR (100 MHz, DMSO-d_6, δ): 79.40, 83.64, 114.99, 117.95, 118.15, 121.97 (q, J = 32 Hz), 124.49 (q, J = 270 Hz), 126.02 (q, J = 3.5 Hz), 132.43, 139.98, 143.21, 152.04. Anal. Calcd for $C_{16}H_{11}N_2OF_3$ (304.27): C, 63.16; H, 3.64; N, 9.21. Found: C, 62.96; H, 3.76; N, 9.06.

2.2.9 *Synthesis of 1-(4-Ethynylphenyl)-1-methyl-3-[4-(trifluoromethyl)phenyl]urea (3g)*

4-(Trifluoromethyl)phenyl isocyanate (0.69 mL, 4.91 mmol) was added to a solution of *N*-methyl-4-ethynylaniline (0.57 g, 4.35 mmol) in dry THF (20 mL) at room temperature under an N_2 atmosphere. The reaction mixture was then stirred for 5 h. After removal of the solvent, the residue was dissolved in ethyl acetate and washed with 1 mol L^{-1} HCl, followed by water. The organic layer was dried over anhydrous Na_2SO_4 and evaporated to dryness. The residue was purified by column chromatography on silica gel with dichloromethane to provide **3g** as a solid. Yield = 0.86 g (63 %). ^{1}H NMR (DMSO-d_6, 400 MHz, δ): 3.30 (s, 3H, –N–CH_3), 4.20 (s, 1H, $HC \equiv$ C-), 7.34 (d, J = 8.6 Hz, 2H, aromatic), 7.50 (d, J = 8.5 Hz, 2H, aromatic), 7.58 (d, J = 8.8 Hz, 2H, aromatic), 7.67 (d, J = 8.8 Hz, 2H, aromatic), 8.80 (s, 1H, –N*H*–CO–). ^{13}C NMR (100 MHz, DMSO-d_6, δ): 37.25, 80.67, 83.16, 118.75, 119.29, 122.01 (q, J = 32 Hz), 124.54 (q, J = 270 Hz), 125.99 (q, J = 4.1 Hz), 132.57, 143.87, 144.18, 154.20. Anal. Calcd for $C_{17}H_{13}N_2OF_3$ (318.29): C, 64.15; H, 4.12; N, 8.80; Found: C, 64.23; H, 4.13; N, 8.84.

2.2.10 Synthesis of Azido End-Functionalized PNIPAM (2) by the ATRP

NIPAM (50.0 g, 442 mmol) and CuCl (1.75 g, 17.7 mmol) were added to a round bottom flask. The flask was capped with a septum, and purged with argon. Degassed 2-propanol (116 mL), Me_6TREN in degassed 2-propanol (6.00 mL, 17.7 mmol, 2.95 mol L^{-1}), and **1** in degassed 2-propanol (6.00 mL, 17.7 mmol, 2.95 mol L^{-1}) were then sequentially added. The reaction mixture was stirred for 4 h at room temperature. The polymerization was quenched by exposure to air. Aliquots were removed from the reaction mixture to determine the conversion of NIPAM from the ^{1}H NMR spectrum (97.1 %). The reaction mixture was diluted with THF and passed through a short column of silica gel to remove the copper complex. The obtained solution was dialyzed using a cellophane tube (Spectra/Por® 6 Membrane; MWCO: 1000) in methanol and finally reprecipitated from THF into hexane to provide the azido end-functionalized PNIPAM (**2**) as a white solid. Yield = 23.8 g (45 %). $M_{n,NMR}$ = 3.3 kg mol^{-1}, $M_{n,SEC}$ = 5.7 kg mol^{-1}, M_w/M_n = 1.19.

2.2.11 The CuAAC of 2 and Alkyne Compounds for the Synthesis of End-Functionalized PNIPAMs (4a–h)

A typical procedure for the CuAAC is as follows, a solution of CuCl (70.2 mg, 0.708 mmol) and PMDETA (296 μL, 1.42 mmol) in degassed THF (0.50 mL) was added to a solution of **2** (600 mg, 0.182 mmol) and **3f** (215 mg, 0.708 mmol) in degassed THF (7.00 mL) under an argon atmosphere. After stirring for 7 h at room temperature, the reaction mixture was diluted with THF and then passed through a short column of silica gel to remove the copper complex. The obtained solution was dialyzed using a cellophane tube (Spectra/Por® 6 Membrane; MWCO: 1000) in methanol and finally freeze-dried from water to give the 4-(trifluoromethyl) phenyl urea end-functionalized PNIPAM (**4f**) as a white solid. Yield = 322 mg

(54 %). $M_{n,SEC} = 8.9$ kg mol^{-1}, $M_w/M_n = 1.17$. The reactions using other alkynes were carried out with the same procedures.

2.2.12 Synthesis of Triazolyl End-Functionalized PNIPAM (4i) from 4h

TBAF in THF (1.50 mL, 1.50 mmol, 1.0 mol L^{-1}) was added to a solution of trimethylsilyl end-functionalized PNIPAM (**4h**) (182 mg, 55.2 μmol) in dry THF (1.25 mL) and methanol (1.00 mL) at room temperature under an N_2 atmosphere [29]. After stirring overnight, the reaction mixture was diluted with THF then passed through a short column of silica gel. The obtained solution was dialyzed using a cellophane tube (Spectra/Por® 6 Membrane; MWCO: 1000) in methanol followed by water and finally freeze-dried to give the polymer **4i** as a white solid. Yield = 51.4 mg (29 %). $M_{n,SEC} = 5.9$ kg mol^{-1}, $M_w/M_n = 1.22$.

2.2.13 Turbidimetric Analysis

A typical procedure for the turbidimetric analysis is as follows: an aqueous solution of a polymer sample (2.0 mg mL^{-1}) was prepared and cooled in an ice bath for 2 min. The resulting clear solution was then transferred to a poly(methyl methacrylate) cell with a 1-cm path length. The transmittance at 500 nm of the aqueous solution was recorded by a UV-vis spectrophotometer equipped with a temperature controller. The solution temperature was gradually increased at the heating rate of 1.0 °C min^{-1}.

2.2.14 Dynamic Light Scattering Measurement

The hydrodynamic diameter (D_h) and size distribution for the polymer aggregated in an aqueous solution were determined by dynamic light scattering (DLS) at the fixed scattering angle of 90°. The polymer sample for the measurement was dissolved in deionized water at room temperature to prepare an aqueous solution (0.20 mg mL^{-1}). All samples were first filtered through a 0.45 μm membrane filter. The measurement was carried out after the sample stood for at least 1 h at 25 °C. The D_h and particle size distribution were calculated by the CONTIN analysis.

2.2.15 Transmission Electron Microscopy Measurement

To prepare a sample for the TEM measurement, an aqueous solution of **4f** (0.20 mg mL^{-1}) was stirred for 72 h at room temperature in order to reach thermodynamic equilibrium. A drop of the polymer solution was then deposited onto a

200-mesh carbon-coated copper grid and dried overnight in a desiccator to achieve dryness. The TEM measurement was performed on the obtained polymer sample at the acceleration voltage of 100 kV.

2.3 Results and Discussion

2.3.1 Synthesis of Diphenylurea End-Functionalized PNIPAMs

In this study, the author attempted to compare the thermoresponsive properties of a series of urea end-functionalized PNIPAMs with the same number-average degree of polymerization in order to accurately evaluate the effect of the terminal urea groups on the thermoresponsive properties of PNIPAM by excluding interference from other factors. Thus, the synthesis of the urea end-functionalized PNIPAMs was carried out by following the synthetic strategy for end-functionalized PNIPAMs previously demonstrated by Kakuchi et al., which is based on (i) the synthesis of azido end-functionalized PNIPAM by the atom transfer radical polymerization (ATRP) of NIPAM using an initiator bearing an azido group, and (ii) the CuAAC reaction between the resultant azido end-functionalized PNIPAM and an alkyne compound, as shown in Scheme 2.1 [18]. In principle, a variety of end-functionalized PNIPAMs with the same number-average degree of polymerization is obtained when the same azido end-functionalized PNIPAM is used as the starting material. First, the ATRP of NIPAM was carried out using *N*-(2-azidoethyl)-2-chloropropionamide (**1**) as the initiator and CuCl/Me_6TREN as the catalyst system to afford the azido end-functionalized PNIPAM (**2**), which was a precursor polymer for further urea conjugation. The number-average molecular weight determined by a ^{1}H NMR measurement ($M_{n,NMR}$) of the resulting product was 3.3 kg mol^{-1} and in good agreement with the theoretical value of 2.9 kg mol^{-1}. The number-average degree of polymerization for the obtained **2** was determined to be 27.6 from the $M_{n,NMR}$. The polydispersity index (M_w/M_n) of the obtained **2** was estimated to be 1.19 as the result of the SEC measurement in DMF containing 0.01 mol L^{-1} of LiCl using narrow-dispersed polystyrene as the calibration standard. Subsequently, the CuAAC between **2** and various diphenylurea derivatives (**3a–f**) was carried out in THF at room temperature using the CuCl/PMDETA complex as the catalyst system. **3a–f** were synthesized from 4-ethynylaniline and various isocyanates. Completion of the reaction was confirmed by the ^{1}H NMR and IR analyses. Figures 2.1 and 2.2 show the change in the NMR and IR spectra of **2** and **3f** as an example. The signal due to the two methylene groups at the chain end of **2** at 3.49 ppm, signals *g* and *f*, was not observed in the ^{1}H NMR spectrum of **4f**, as shown in Fig. 2.1. Instead, those of the methylene group adjacent to the triazolyl group, signal h, as well as the aromatic groups, signal aromatic, were observed at 4.6 ppm and 7.4–8.2 ppm, respectively. The absorption of the azido group originally observed at 2,102 cm^{-1} in the IR

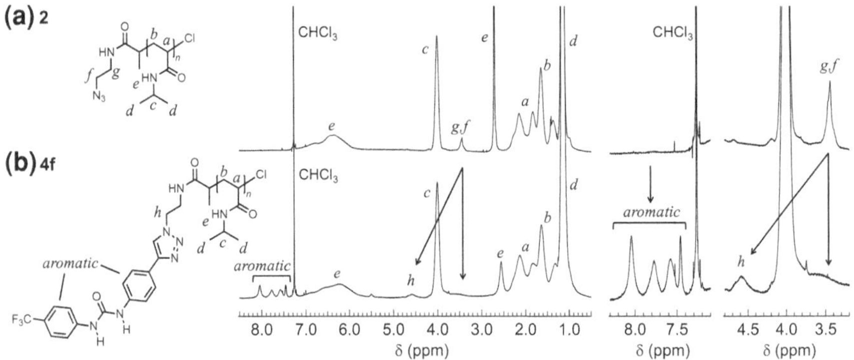

Fig. 2.1 ^{1}H NMR spectra of **a 2** and **b 4f** in $CDCl_3$

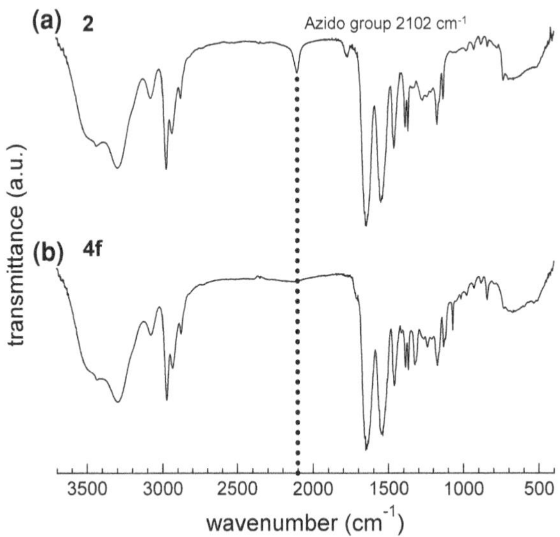

Fig. 2.2 IR spectra of **a 2** and **b 4f**

spectrum of **2** disappeared in the spectrum of **4f**, as shown in Fig. 2.2. The same change was observed for the obtained **3a–e**. Thus, the precise synthesis of **3a–f** was achieved by the combination of the ATRP and the CuAAC.

Moreover, the CuAAC of **2** and **3g**, namely, the *N*-methylated analogue of **3f**, was carried out to produce **4g** in order to elucidate the contribution of the hydrogen bonding ability of the terminal urea groups to the change in the thermoresponsive properties of PNIPAM. The hydrogen bonding ability of the terminal urea group in **4g** was expected to be weaker than that of **4f**. Completion of the reactions was confirmed by the ^{1}H NMR and IR analyses. Furthermore, the triazolyl end-functionalized PNIPAM (**4i**) as the control polymer was synthesized by the CuAAC between **2** and trimethylsilylacetylene followed by desilylation with TBAF in

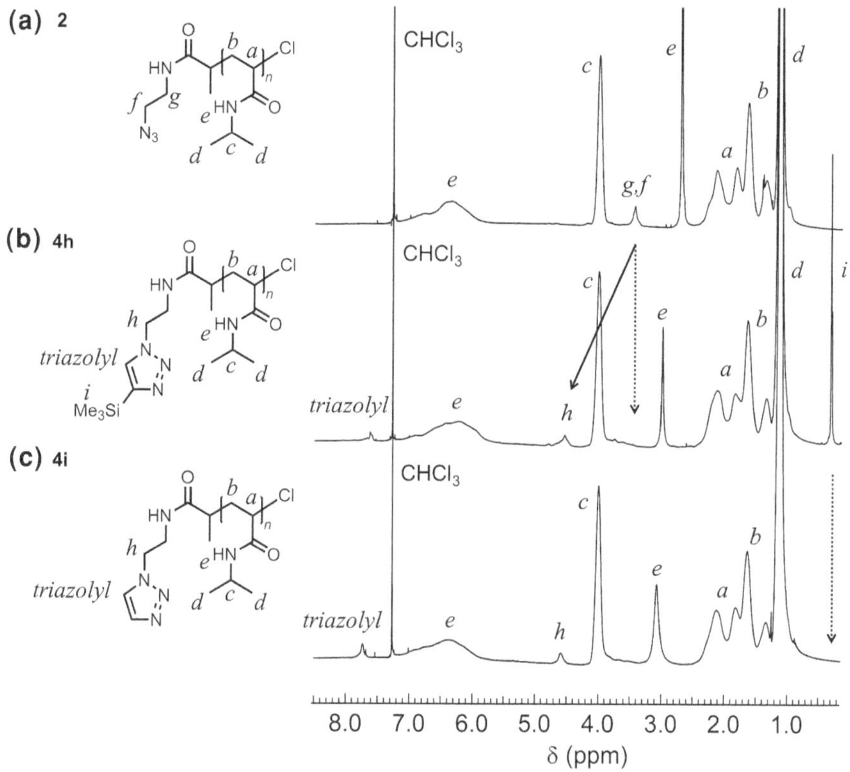

Fig. 2.3 ^{1}H NMR spectra of **2** (**a**), **4h** (**b**), and **4i c** in $CDCl_3$

order to evaluate the effect of the diphenylurea moiety on the thermoresponsive properties of the urea end-functionalized PNIPAMs. Completion of the two reactions was confirmed by the ^{1}H NMR analyses, as shown in Fig. 2.3. The signal due to the two methylene groups at the chain end of **2** disappeared in the ^{1}H NMR spectrum of **4h**. Instead, the signal due to the triazolyl group, signal *triazolyl*, and the trimethylsilyl group, signal *i*, were observed at 7.6 and 0.3 ppm, respectively. The signal due to the trimethylsilyl group disappeared in the spectrum of **4i**. In summary, urea end-functionalized PNIPAMs (**4a–i**) with the same number-average degree of polymerization were successfully synthesized by the CuAAC between the PNIPAM precursor **2** and various urea derivatives.

2.3.2 *Effect of the Terminal Urea Group on the Thermoresponsive Properties of PNIPAM*

The thermoresponsive properties of a series of PNIPAMs with different terminal structures were characterized by a turbidimetric analysis, as shown in Fig. 2.4. Every end-functionalized PNIPAM showed the characteristic phase transition

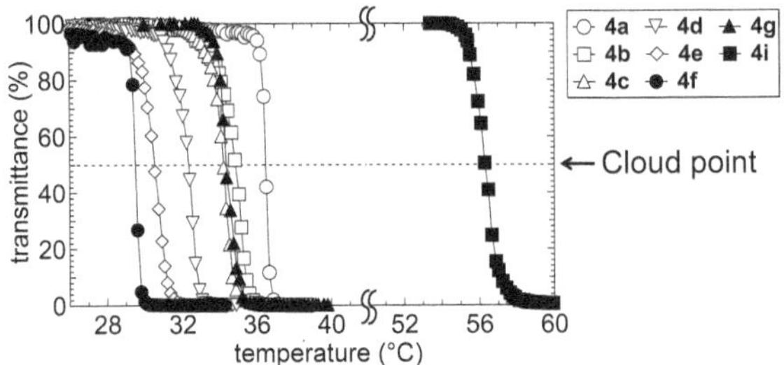

Fig. 2.4 Transmittance ($\lambda = 500$ nm) versus temperature for aqueous solutions of **4a–g** and **i** (2.0 mg mL^{-1})

Table 2.1 Cloud points of the obtained end-functionalized PNIPAMs (**4a–g** and **i**)

Polymer	Terminal structure	Cloud point (°C)[a]
4a		36.5
4b		34.8
4c		34.3
4d		32.4
4e		30.6
4f		29.5
4 g		34.4
4i		56.3

[a]Determined by turbidimetric analysis
Conditions: concentration = 2.0 mg mL^{-1}, heating rate = 1.0 °C min^{-1}, cloud point was determined by the temperature which the transmittance ($\lambda = 500$ nm) of aqueous solution reach 50 %

phenomenon at a temperature specific to each polymer sample. The cloud point was defined as the temperature at which the transmittance of the sample solution reached 50 %. Table 2.1 summarizes the cloud points of all the polymers. The cloud points of **4a–f** ranged from 29.5 to 36.5 °C, whereas **4i**, the control polymer, had a cloud point at 56.3 °C. Thus, the introduction of the diphenylurea group to the chain end of PNIPAM caused more than a 20 °C change in the drastic decrease of the cloud point.

A significant variation in the cloud point was observed for **4a–f**, though only the substituent group in the terminal diphenylurea group is the difference among the polymer structures. **4d**, **4e**, and **4f** bearing an electron withdrawing group, namely a nitro group, a chlorine atom, and a trifluoromethyl group, in the urea group showed cloud points at 32.4, 30.6, and 29.5 °C, respectively. In contrast, higher cloud points of 34.8 and 34.3 °C were observed for **4b** and **c** bearing an electron donating group, namely a methoxy group and a methyl group, in the urea group. Therefore, the variation in the cloud point was found to obviously correlate with the substituent effect, though the exceptionally high cloud point of **4a**, 36.5 °C, cannot be explained only by such an effect. Given that the substituent effect of the aromatic ring directly influences the electronic properties of the neighboring urea group, these results suggest that such variations in the cloud point should be ascribable to the hydrogen bonding ability of the terminal urea functionality.

In order to provide further insight into the relationship between the hydrogen bonding ability and the cloud point, the thermoresponsive behavior of **4g**, of which the hydrogen bonding ability was essentially inhibited, was examined. Despite the very small difference in the structure, the cloud point of **4g** was 5 °C higher than that of **4f**. The cloud point of **4g** should be lower than that of **4f** or at least unchanged if the cloud point varies only by the hydrophobicity of the polymer structure [17, 18]. Thus, this comparison well demonstrated the importance of the hydrogen bonding ability on the variation in the cloud point. Based on all the results regarding the thermoresponsive properties, the hydrogen bonds derived from the terminal urea groups can be concluded to efficiently work even in water, leading to a decrease in the cloud point.

2.3.3 Aggregation State of Urea End-Functionalized PNIPAM in Water

In the previous section, the robust and efficient hydrogen bonding was revealed to be actually present even in water for the series of urea end-functionalized PNIPAMs. Given that hydrogen bonding of the urea groups realizes a continuous and hierarchical intermolecular interaction, the polymers should self-assemble in water even at a temperature below the cloud point. Thus, the ^{1}H NMR measurements of **4f** and **g** were carried out in D_2O at 25 °C to elucidate the aggregation behavior of the polymers, as shown in Fig. 2.5. No significant difference was observed in the signals originated from the main chain protons of **4f** and **g**. Moreover, a direct investigation of the urea protons was unavailable because of the rapid exchange with the deuterium atoms of D_2O. An obvious difference was observed in the signals due to the aromatic protons at the chain end, which appeared between 6.8 and 8.0 ppm. The signals due to the aromatic protons of **4f** were broad, while those of **4g** were observed as sharp peaks. Therefore, the peak broadening was found to closely correlate with the hydrogen bonding ability of the

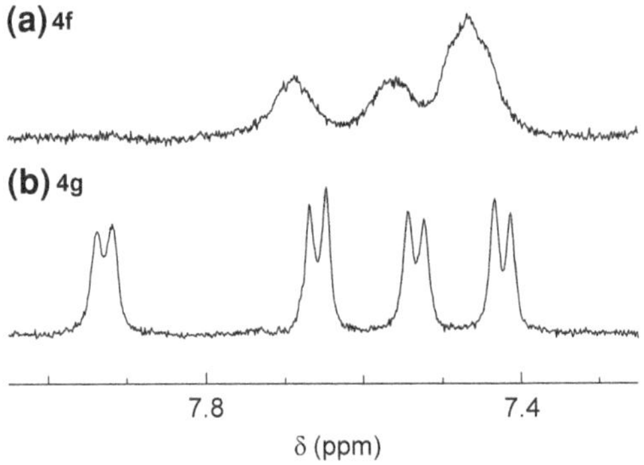

Fig. 2.5 ^{1}H NMR spectra of (**a**) **4f** and (**b**) **4g** in D_2O (2.0 mg mL^{-1}) at 25 °C

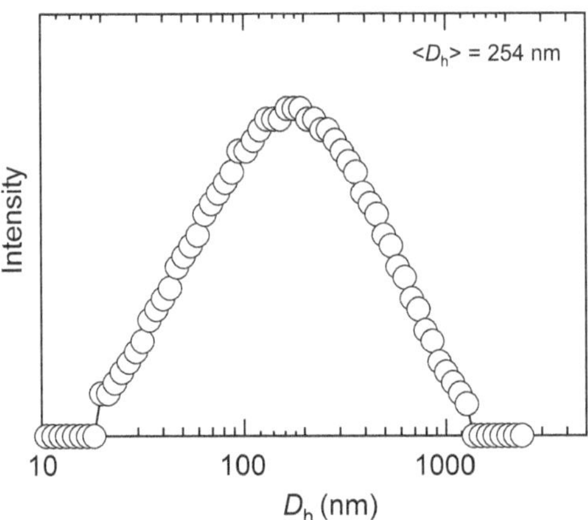

Fig. 2.6 Intensity weighted size distributions of **4f** in water (0.20 mg mL^{-1}) at 25 °C obtained by the DLS measurement using the CONTIN analysis

urea moiety. These results indicate that the mobility of the chain ends for **4f** was probably restricted due to the strong hydrogen bonding ability of terminal urea group, implying that the polymer chains assemble via intermolecular hydrogen bonding even at a temperature below the cloud point.

In order to prove the existence of the polymer assemblies in water, DLS and TEM measurements were carried out for **4f** in water, which was considered to have the highest hydrogen bonding ability in the series of urea end-functionalized PNIPAMs. The DLS measurement was conducted using aqueous solutions of **4f** (0.20 mg mL^{-1}) at 25 °C. The result was analyzed using the CONTIN method, and Fig. 2.6 depicts the obtained particle size distribution. From the DLS analysis, **4f** was found to actually form the aggregation under this condition, of which the average hydrodynamic diameter ($\langle D_h \rangle$) was determined to be 254 nm. It should be noted that the thermoresponsive behavior of PNIPAM itself is excluded from the driving force for such an aggregation process because the measurement was carried out at a temperature below the cloud point of **4f**. Thus, this result indicates that the strong hydrogen bonding ability of the terminal urea group on **4f** enabled the formation of the aggregates in water. However, the broad size distribution suggested a multiplicity of the aggregation morphology.

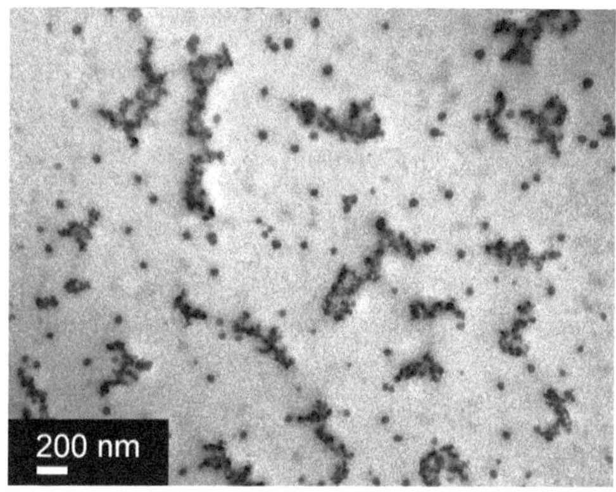

Fig. 2.7 TEM micrograph of aggregates in the aqueous solution of **4f** (0.20 mg mL^{-1}) at room temperature

For the direct observation of the aggregate of **4f** in water, the TEM measurement was carried out, in which the sample was prepared from an aqueous solution of **4f** (0.20 mg mL^{-1}) at room temperature. The resulting TEM image showed spherical nanoparticles distributed over the entire area, as shown in Fig. 2.7. Each particle was considered to be an aggregate of **4f**, of which the diameter ranged from 35 to 45 nm. Interestingly, part of the nanoparticles seemed to further assemble and form larger aggregates with diameters greater than 100 nm, which should correspond to the bigger scatterers observed in the DLS measurement. These results substantiated that **4f** supramolecularly assembled in water through the intermolecular hydrogen bonding of the chain end urea groups.

2.4 Conclusions

The author has demonstrated an adjustment of the thermoresponsive properties of PNIPAM through the hydrogen bonding in water by introducing a series of diphenylurea groups to the chain end. The combination of the ATRP and the CuAAC was employed for preparing the urea end-functionalized PNIPAMs with the same number-average degree of polymerization and different substituent groups in the urea group, which allowed evaluating the effect of the introduction of the urea group on the thermoresponsive properties of PNIPAM while excluding all other factors that influence the thermoresponsive properties. For the series of urea end-functionalized PNIPAMs, the intermolecular hydrogen bonding of the chain end urea group was revealed to work even in water, leading to the self-assembly of the polymer chains at a temperature below the cloud point. Such an antecedent aggregation made it possible to facilitate the phase transition process, resulting in the drastic decrease of the cloud point. Thus, the present study realized the control of the macromolecular function by the hydrogen bonding formation in water for the first time.

References

1. Paleos CM, Tsiourvas D (1997) Molecular recognition of organized assemblies via hydrogen bonding in aqueous media. Adv Mater 9:695–710
2. Fenniri H, Packiarajan M, Vidale KL, Sherman DM, Hallenga K, Wood KV, Stowell JG (2001) Helical rosette nanotubes: Design, self-assembly, and characterization. J Am Chem Soc 123:3854–3855
3. Jung JH, John G, Masuda M, Yoshida K, Shinkai S, Shimizu T (2001) Self-assembly of a sugar-based gelator in water: Its remarkable diversity in gelation ability and aggregate structure. Langmuir 17:7229–7232
4. Brunsveld L, Vekemans J, Hirschberg J, Sijbesma RP, Meijer EW (2002) Hierarchical formation of helical supramolecular polymers via stacking of hydrogen-bonded pairs in water. Proc Natl Acad Sci USA 99:4977–4982
5. Kawasaki T, Tokuhiro M, Kimizuka N, Kunitake T (2001) Hierarchical self-assembly of chiral complementary hydrogen-bond networks in water: Reconstitution of supramolecular membranes. J Am Chem Soc 123:6792–6800
6. Chebotareva N, Bomans PHH, Frederik PM, Sommerdijk N, Sijbesma RP (2005) Morphological control and molecular recognition by bis-urea hydrogen bonding in micelles of amphiphilic tri-block copolymers. Chem Commun 4967–4969
7. Schild HG (1992) Poly(*N*-isopropylacrylamide): experiment, theory and application. Prog Polym Sci 17:163–249
8. Coughlan DC, Quilty FP, Corrigan OI (2004) Effect of drug physicochemical properties on welling/deswelling kinetics and pulsatile drug release from thermoresponsive poly(*N*-isopropylacrylamide) hydrogelsJ Control Rel 98:97–114
9. Tsuda Y, Yamato M, Kikuchi A, Watanabe M, Chen GP, Takahashi Y, Okano T (2007) Thermoresponsive microtextured culture surfaces facilitate fabrication of capillary networks. Adv Mater 19:3633–3636
10. De P, Li M, Gondi SR, Sumerlin BS (2008) Temperature-regulated activity of responsive polymer−protein conjugates prepared by grafting-from via RAFT polymerization. J Am Chem Soc 130:11288–11289
11. Klouda L, Mikos AG (2008) Thermoresponsive hydrogels in biomedical applications - a review. Eur J Pharm Biopharm 68:34–45
12. Ito T, Hioki T, Yamaguchi T, Shinbo T, Nakao S, Kimura S (2002) Development of a molecular recognition ion gating membrane and estimation of its pore size control. J Am Chem Soc 124:7840–7846
13. Nykanen A, Nuopponen M, Laukkanen A, Hirvonen SP, Rytela M, Turunen O, Tenhu H, Mezzenga R, Ikkala O, Ruokolainen J (2007) Phase Behavior and temperature-responsive molecular filters based on self-assembly of polystyrene-*block*-poly(*N*-isopropylacrylamide)-*block*-polystyrene. Macromolecules 40:5827–5834
14. Tokuyama H, Iwama T (2007) Temperature-swing solid-phase extraction of heavy metals on a poly(*N*-isopropylacrylamide) hydrogel. Langmuir 23:13104–13108
15. Wang H, An YL, Huang N, Ma RJ, Li JB, Shi LQ (2008) Contractive polymeric complex micelles as thermo-sensitive nanopumps. Macromol Rapid Commun 29:1410–1414
16. Kujawa P, Segui F, Shaban S, Diab C, Okada Y, Tanaka F, Winnik FM (2006) Impact of end-group association and main-chain hydration on the thermosensitive properties of hydrophobically modified telechelic poly(*N*-isopropylacrylamides) in water. Macromolecules 39:341–348
17. Xia Y, Burke NAD, Stöver HDH (2006) End group effect on the thermal response of narrow-disperse poly(*N*-isopropylacrylamide) prepared by atom transfer radical polymerization. Macromolecules 39:2275–2283
18. Narumi A, Fuchise K, Kakuchi R, Toda A, Satoh T, Kawaguchi S, Sugiyama K, Hirao A, Kakuchi T (2008) A versatile method for adjusting thermoresponsivity: synthesis and 'click' reaction of an azido end-functionalized poly(*N*-isopropylacrylamide). Macromol Rapid Commun 29:1126–1133

19. Ray B, Isobe Y, Matsumoto K, Habaue S, Okamoto Y, Kamigaito M, Sawamoto M (2004) Macromolecules 37:1702–1710
20. Ray B, Okamoto Y, Kamigaito N, Sawamoto M, Seno K, Kanaoka S, Aoshima S (2005) Effect of tacticity of poly(*N*-isopropylacrylamide) on the phase separation temperature of its aqueous solutions. Polym J 37:234–237
21. Hirano T, Okumura Y, Kitajima H, Seno M, Sato T (2006) Dual roles of alkyl alcohols as syndiotactic-specificity inducers and accelerators in the radical polymerization of *N*-isopropylacrylamide and some properties of syndiotactic poly(*N*-isopropylacrylamide). J Polym Sci, Part A: Polym Chem 44:4450–4460
22. Katsumoto Y, Kubosaki N (2008) Tacticity effects on the phase diagram for poly(*N*-isopropylacrylamide) in water. Macromolecules 41:5955–5956
23. Duan Q, Miura Y, Narumi A, Shen X, Sato S, Satoh T, Kakuchi T (2006) Synthesis and thermoresponsive property of end-functionalized poly(*N*-isopropylacrylamide) with pyrenyl group. J Polym Sci, Part A: Polym Chem 44:1117–1124
24. Duan Q, Narumi A, Miura Y, Shen XD, Sato S-I, Satoh T, Kakuchi T (2006) Thermoresponsive property controlled by end-functionalization of poly(*N*-isopropylacrylamide) with phenyl, biphenyl, and triphenyl groups. Polym J 38:306–310
25. Furyk S, Zhang YJ, Ortiz-Acosta D, Cremer PS, Bergbreiter DE (2006) Effects of end group polarity and molecular weight on the lower critical solution temperature of poly(*N*-isopropylacrylamide). J Polym Sci, Part A: Polym Chem 44:1492–1501
26. Kujawa P, Tanaka F, Winnik FM (2006) Temperature-dependent properties of telechelic hydrophobically modified poly(*N*-isopropylacrylamides) in water: Evidence from light scattering and fluorescence spectroscopy for the formation of stable mesoglobules at elevated temperatures. Macromolecules 39:3048–3055
27. Obert E, Bellot M, Bouteiller L, Andrioletti F, Lehen-Ferrenbach C, Boué F (2007) Both water- and organo-soluble supramolecular polymer stabilized by hydrogen-bonding and hydrophobic interactions. J Am Chem Soc 129:15601–15605
28. Qu WC, Kung MP, Hou C, Oya S, Kung HF (2007) Quick assembly of 1,4-diphenyltriazoles as probes targeting beta-amyloid aggregates in Alzheimer's disease. J Med Chem 50:3380–3387
29. Chan DCM, Laughton CA, Queener SF, Stevens MFG (2002) Structural studies on bioactive compounds. part 36: design, synthesis and biological evaluation of pyrimethamine-based antifolates against *Pneumocystis carinii*. Bioorg Med Chem 10:3001–3011

Chapter 3
Precise Synthesis of Poly(*N,N*-Dimethylacrylamide) by Group Transfer Polymerization Using a Strong Brønsted Acid and an Amino Silyl Enolate

3.1 Introduction

The controlled/living polymerization has been expanding its use involving a monomer, an initiator, a terminator, and polymerization mechanisms due to increasing demand for the precise synthesis of well-defined macromolecular architectures since the discovery of the living anionic polymerization of styrene by Szwarc in 1956. For polar vinyl monomers, such as the (meth)acrylates and (meth)acrylamides, the applicable scope of the controlled/living radical polymerization is generally broader than that of the anionic one. The controlled/living polymerizations of acrylamide monomers have been achieved by both of the polymerizations, which have realized the synthesis of thermoresponsive polyacrylamides with well-defined structures. However, the unavoidable side reactions in the controlled/living radical polymerizations, such as the radical-radical coupling and the disproportionation, can cause defects in the polymer structure and broadening of the molecular weight distribution. The anionic polymerization required extensive purification of the reagents and special apparatus for the polymerization, though it provides excellent molecular weight control. Thus, the author focused on the development of the group transfer polymerization (GTP), which is one of the important methods of the living anionic polymerization [1–3]. GTP is principally free from any termination reactions to provide a well-defined polymer and can be conducted under milder condition than the anionic polymerization. However, the GTP has been unsuitable for the polymerization of *N,N*-dialkyl(meth)acrylamides though Ishizone and Nakahama et al. reported the living anionic polymerization of a series of *N,N*-dialkyl(meth)acrylamides [4–10]. For example, Sogah and Webster et al. [11] reported that the GTP of *N,N*-dimethylacrylamide (DMAA) afforded low molecular weight polymers with a relatively broad polydispersity index (>1.62). Freitag et al. [12–14] reported that the GTP of *N,N*-diethylacrylamide produced the poly(*N,N*-diethylacrylamide) with less than a 3 kg mol^{-1} molecular weight, though its polydispersity index was as low as 1.12–1.19. Hence, the GTP of *N,N*-dialkyl(meth)acrylamide leading to

K. Fuchise, *Design and Precise Synthesis of Thermoresponsive Polyacrylamides*, Springer Theses, DOI: 10.1007/978-4-431-55046-4_3, © Springer Japan 2014

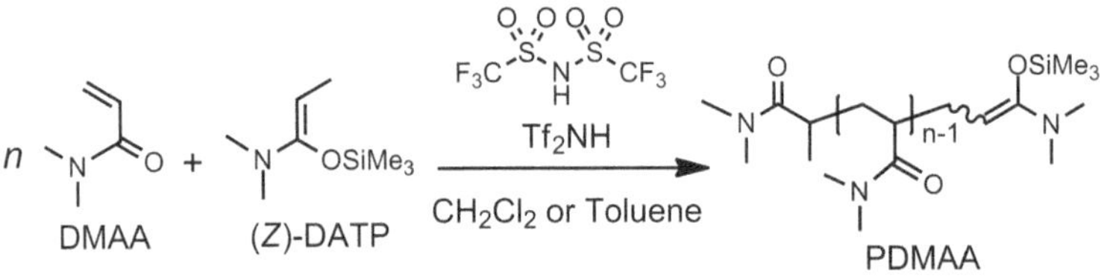

Scheme 3.1 GTP of *N*,*N*-dimethylacrylamide (DMAA) promoted by Tf_2NH using an amino silyl enolate as the initiator (Adapted with permission from Ref. [26]. Copyright 2010 American Chemical Society)

well-defined polymers, especially the thermoresponsive polyacrylamides, still remains a challenging task.

Recently, organocatalytic polymerization has turned into one of the common methods for preparing well-defined polymers, which began not only to improve well-established polymerization systems, but also to provide polymer chemistry with new concepts and polymerization systems. For example, Hedrick et al. reported that the *N*-heterocyclic carbene (NHC) mediated ring-expansion polymerization of cyclic esters led to the spontaneous formation of cyclic polyesters with predictable molecular weights and narrow polydispersity indices [15]. Furthermore, Taton and co-workers reported a smart synthetic strategy for the α,ω-hetero-end-functionalized poly(ethylene oxide) using the zwitterionic ring-opening polymerization of ethylene oxide with NHC as the organocatalyst [16]. In the growing field of organocatalytic polymerizations, the organocatalyst was also found to be a new and versatile catalyst for the GTP of (meth)acrylates; Taton et al. and Hedrick et al. independently reported that NHC successfully catalyzed the GTP of (meth)acrylates using 1-methoxy-1-trimethylsiloxy-2-methyl-1-propene (MTS) as the initiator to produce well-defined homo- and block polymers [17–19]. In addition, Kakuchi et al. [20] recently reported that bis(trifluoromethanesulfonyl)imide (Tf_2NH) [21–25], a very strong Brønsted acid, is capable of promoting the GTP of methyl methacrylate (MMA) using MTS as the initiator to afford the highly syndiotactic poly(methyl methacrylate) without any obvious side reactions. Of significant importance, the catalytic activity of Tf_2NH in the GTP process was significantly higher than the conventional Lewis acids, which suggested that the Tf_2NH has the potential to realize the GTP of *N*,*N*-dialkyl(meth)acrylamides proceeding in a living fashion by the appropriate tuning of the polymerization conditions.

In this chapter, the author describes the GTP of DMAA using Tf_2NH, as shown in Scheme 3.1. (*Z*)-1-Dimethylamino-1-trimethylsiloxy-1-propene ((*Z*)-DATP) was synthesized as an initiator for the polymerization to compare its initiating performance to that of MTS. The living nature of the Tf_2NH-promoted GTP of DMAA was assessed by the molecular weight control of the resultant poly(*N*,*N*-dimethylacrylamide) (PDMAA), kinetic study, postpolymerization experiment, and mass spectral analysis of the obtained PDMAA. The mechanism of the polymerization was briefly investigated. Lastly, the stereoregularity and the thermal properties of the obtained PDMAA were analyzed in detail.

3.2 Experimental Section

3.2.1 Materials

N,N-Dimethylacrylamide (DMAA), *n*-butyl lithium (1.6 mol L^{-1} in *n*-hexane), dry dichloromethane (CH_2Cl_2, >99.5 %; water content, <0.001 %), dry acetonitrile (MeCN, >99.5 %; water content, <0.002 %), tetrahydrofuran (THF), toluene, 2-propanol, and pyridine were purchased from Kanto Chemicals Co., Inc. DMAA was dried over CaH_2 and distilled two times under reduced pressure. CH_2Cl_2 was distilled from CaH_2 and degassed by three freeze-pump-thaw cycles. THF and toluene were distilled from sodium benzophenone ketyl. 1-Methoxy-1-trimethylsiloxy-2-methyl-1-propene (MTS), *N*-(trimethylsilyl) bis(trifluoromethanesulfonyl)imide (Tf_2NSiMe_3), diisopropylamine, *N,N*-dimethylpropionamide, chlorotrimethylsilane, and *trans*-3-indoleacrylic acid were purchased from Tokyo Kasei Kogyo Co., Ltd. MTS was purified by fractional distillation under argon. Diisopropylamine, *N,N*-dimethylpropionamide, and chlorotrimethylsilane were dried over CaH_2 followed by distillation. Tf_2NH, tris(dimethylamino)sulfonium difluorotrimethylsilicate (TAS-$SiMe_3F_2$), tetra-*n*-butylammonium acetate (TBA-AcO), and sodium trifluoroacetate were available from the Sigma-Aldrich Chemicals Co. and used as received. Zinc iodide (ZnI_2) was purchased from Junsei Chemical Co., Ltd., and used as received. Insulin (bovine) was purchased from TAKARA BIO, Inc. and used as received. Tris(dimethylamino)sulfonium bifluoride (TAS-HF_2) was prepared from TAS-$SiMe_3F_2$ according to the method reported by Webster and co-workers [11]. All other chemicals were purchased from various suppliers and used without purification.

3.2.2 Measurements

The ^{1}H (400 MHz) and ^{13}C NMR (100 MHz) spectra were recorded using a JEOL JNM-A400II and JEOL-ECP400. The preparation of the polymerization solution was carried out in an MBRAUN stainless steel glovebox equipped with a gas purification system (molecular sieves and copper catalyst) in a dry argon atmosphere (H_2O, O_2 <1 ppm). The moisture and oxygen contents in the glovebox were monitored by an MB-MO-SE 1 and an MB-OX-SE 1, respectively. The characterizations for the molecular weights of PDMAA were performed by size exclusion chromatography (SEC) (pump: Jasco PU-2080 Plus, degasser: Jasco 2080-53, column oven: Jasco CO-2065 Plus, temperature of the column oven: 40 °C) using DMF containing 10 mmol L^{-1} LiBr as the eluent and at the flow rate of 1.0 mL min^{-1} at room temperature (25 °C). The SEC was equipped with three columns (Shodex KD806 M × 2; size: 8 mm × 300 mm, average bead size: 5 μm, exclusion limit: 2×10^7 and Shodex OHpak SB-8025HQ;

size: 8 mm × 300 mm, particle size: 6 μm, exclusion limit: 1 × 10^4), an refractive index detector (RI: Shodex RI-71), and a multiangle laser light scattering detector (MALS; Wyatt Technology DAWN-DSP, wavelength $\lambda = 632.8$ nm). The weight averaged-molecular weight (M_w) of PDMAA was determined by the MALS. The polydispersity index (M_w/M_n) was determined by the RI based on polystyrene standards with the $M_n(M_w/M_n)$s of 2.18 × 10^3 g mol^{-1} (1.08), 4.76 × 10^3 g mol^{-1} (1.08), 8.00 × 10^3 g mol^{-1} (1.05), 1.58 × 10^4 g mol^{-1} (1.05), 4.12 × 10^4 g mol^{-1} (1.07), 1.10 × 10^5 g mol^{-1} (1.04), and 2.28 × 10^6 g mol^{-1} (1.02). The number averaged-molecular weight (M_n) of PDMAA was calculated from M_w / (M_w/M_n). The Rayleigh ratio $R(90)$ at a scattered angle of 90° was based on that of pure toluene at a wavelength of 632.8 nm at 25 °C. The corrections for the sensitivity of 17 detectors at angles of other than 90° and the dead volume of each detector were performed using the scattering intensities of 0.30 wt% DMF containing 10 mmol L^{-1} LiBr solution of a polystyrene standard with $M_w = 4.41 \times 10^4$ g mol^{-1} and $M_w/M_n = 1.07$. The polymer sample solutions with a mass concentration (C_p) of about 5 × 10^{-4} g mL^{-1} were injected using a sample loop of 100 μL to the SEC columns and diluted 10–10^3 times lower than the original C_p in the columns during the separation. Thus, the concentration effect on the M_w value can be ignored.

The specific refractive index increment (d*n*/d*c*) was measured using a differential refractometer (Otsuka Electronics DRM-1021, wavelength $\lambda = 632.8$ nm) at 25 °C. The solutions of PDMAA ($M_w = 1.12 \times 10^4$ g mol^{-1}) in DMF containing 10 mmol L^{-1} LiBr with the C_p of 1.0–5.0 mg mL^{-1} were prepared by a gravimetric method according to the literature [27]. The measured values were plotted as a function of the sample concentrations and the least-square method was applied, providing the d*n*/d*c* of 0.0786 mL mg^{-1} and this value was used for the M_w determinations of all the PDMAA samples.

The matrix-assisted laser desorption/ionization time-of-flight mass spectrometry (MALDI-TOF MS) measurements were performed using an Applied Biosystems Voyager-DE STR-H mass spectrometer with a 25 kV acceleration voltage. The positive ions were detected in the reflector mode (25 kV). A nitrogen laser (337 nm, 3 ns pulse width, 106–107 W cm^{-2}) operating at 3 Hz was used to produce the laser desorption, and 200 shots were summed. The spectra were externally calibrated using insulin (bovine) with a linear calibration. Samples for the MALDI-TOF MS were prepared by mixing the polymer (1.5 mg mL^{-1}, 10 μL), the matrix (*trans*-3-indoleacrylic acid, 10 mg mL^{-1}, 90 μL), and the cationizing agent (sodium trifluoroacetate, 10 mg mL^{-1}, 10 μL) in THF.

The glass transition temperature (T_g) of the polymers was measured by differential scanning calorimetry (DSC) on a Bruker AXS TG-DTA 3100SA equipped with a Bruker AXS CU9440. The samples were first heated to 180 °C at the heating rate of 20 °C min^{-1}, equilibrated at this temperature for 4 min, and then cooled to 0 °C at the cooling rate of 10 °C min^{-1}. After being held at this temperature for 4 min, the samples were reheated to 360 °C at the heating late of 10 °C min^{-1}.

All the T_g values were obtained from the second scan after removing the thermal history.

3.2.3 Synthesis of (Z)-DATP

To a solution of diisopropylamine (26.7 mL, 190 mmol) in THF (170 mL) was dropwise added *n*-butyllithium (121 mL, 1.57 mol L^{-1} in hexane, 190 mmol) at 0 °C under an argon atmosphere, and the mixture was stirred for 30 min at 0 °C. *N,N*-Dimethylpropionamide (18.8 mL, 173 mmol) was added to the solution of lithium diisopropylamide at 0 °C and the reaction mixture was stirred for 1 h at 0 °C. Trimethylsilylchloride (32.8 mL, 260 mmol) was then added to the reaction mixture at 0 °C. After stirring for 1 h at 0 °C, the entire mixture was filtered and the filtrate was distilled under reduced pressure. The reaction was repeated once again because the purity of the product was 85 %. The product of the first reaction (10.4 mL) was added to a solution of lithium diisopropylamide prepared from diisopropylamine (1.71 mL, 12.2 mmol) and *n*-butyllithium (7.77 mL, 1.57 mol L^{-1} in *n*-hexane, 12.2 mmol) in THF (54 mL). After stirring for 1 h at 0 °C, trimethylsilylchloride (2.05 mL, 16.2 mmol) was added. The reaction mixture was directly distilled from the reaction container under reduced pressure to give (*Z*)-DATP as a colorless liquid with a purity of 99 %. The stereochemistry of the product was determined by a NOESY measurement as the *Z*-isomer and Fig. 3.1 shows the measured spectrum. Yield, 6.86 g (22.9 %). b.p., 68–70 °C / 54 mmHg. ^{1}H NMR (400 MHz, $CDCl_3$): δ 0.19 (s, 9H, -Si(CH_3)$_3$), 1.52 (d, $J = 6.59$ Hz, 3H, = CH–CH_3), 2.46 (s, 6H, -N(CH_3)$_2$), 3.65 (q, $J = 6.59$ Hz, 1H, = *CH*–CH_3). ^{13}C NMR (100 MHz, $CDCl_3$): δ 0.2, 10.9, 40.5, 80.5, 154.7.

3.2.4 Polymerization of DMAA

A typical procedure for the polymerization is as follows: to a solution of DMAA (0.991 g, 10.0 mmol) and (*Z*)-DATP (20.5 μL, 0.10 mmol) in toluene (18.91 mL) was added a stock solution (20 μL, 2.0 μmol) of Tf_2NH in CH_2Cl_2 (0.10 mol L^{-1}) at 0 °C. The polymerization was quenched after 3 h by the addition of a small amount of 2-propanol and pyridine. Aliquots were removed from the reaction mixture to determine the conversion of DMAA from the ^{1}H NMR spectrum. The reaction mixture was dialyzed using a cellophane tube (Spectra/Por® 6 Membrane; MWCO: 1 000) in methanol and freeze-dried from water to provide the polymer as a white solid. When the molecular weight of the obtained polymer was predicted to be near the cutoff molecular weight of the cellophane tube, reprecipitation from ethanol into hexane was carried out instead of dialysis as an intermediate procedure for the purification of the obtained polymer. Yield, 953 mg (96.2 %); $M_n = 11.9$ kg mol^{-1}, $M_w/M_n = 1.11$. The polymerization using TBA-AcO as a

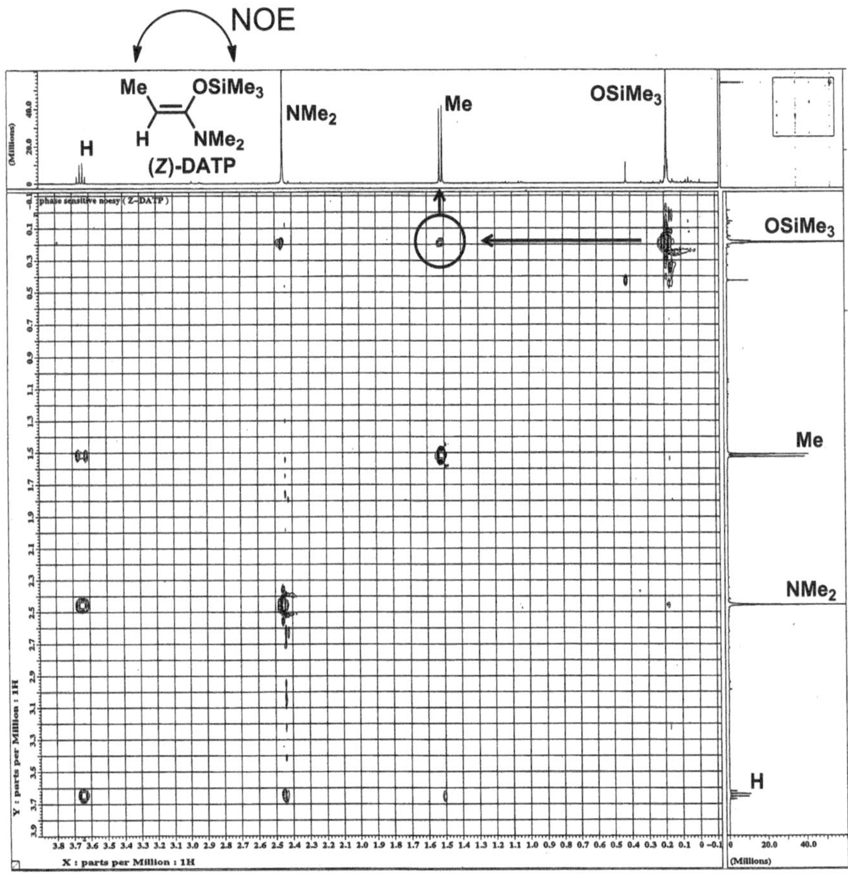

Fig. 3.1 NOESY spectrum of (*Z*)-DATP measured in $CDCl_3$. (Adapted with permission from Ref. [26]. Copyright 2010 American Chemical Society)

catalyst was carried out using the same procedure. In the polymerizations using TAS-HF_2 or ZnI_2 as the catalyst, DMAA was added to a solution of (*Z*)-DATP and a catalyst in toluene to start the polymerization.

3.2.5 Post-polymerization of DMAA

The polymerization of DMAA in toluene at 0 °C was carried out with $[M]_0/[I]_0/[Tf_2NH]_0 = 50/1/0.05$ using the typical procedure for 3 h. Subsequently, the post-polymerization was started by adding 50 equivalents of DMAA to the reaction mixture after aliquots were removed from the reaction mixture for determining the conversion of DMAA and the M_n value of the product. After 4 h, the post-polymerization was quenched by adding a small amount of 2-propanol and pyridine. Following purification, the products were same as the typical polymerization of DMAA.

Table 3.1 Group transfer polymerization (GTP) of *N,N*-dimethylacryamide (DMAA) using Tf_2NH at 0 °C [a] (Adapted with permission from Ref. [26]. Copyright 2010 American Chemical Society)

Run	Initiator	Solvent	$[M]_0/[I]_0/[Tf_2NH]_0$	Time (h)	Conv. (%)[b]	M_n (kg mol^{-1})		M_w/M_n[e]
						Calcd[c]	Obsd.[d]	
1	MTS	CH_2Cl_2	100/1/0.02	2	>99	10.2	20.2	1.24
2	MTS	Toluene	100/1/0.05	3	34.3	3.68	23.7	1.76
3	(*Z*)-DATP	CH_2Cl_2	100/1/0.02	1	>99	10.2	12.7	1.08
4	(*Z*)-DATP	Toluene	100/1/0.05	3	>99	10.5	11.9	1.11
5	(*Z*)-DATP	Toluene	25/1/0.05	3	>99	2.71	3.24	1.16
6	(*Z*)-DATP	Toluene	50/1/0.05	3	>99	5.32	6.10	1.12
7[f]	(*Z*)-DATP	Toluene	200/1/0.05	5	>99	21.0	23.5	1.06
8[g]	(*Z*)-DATP	Toluene	400/1/0.05	9	>99	41.8	53.9	1.07
9	(*Z*)-DATP	Toluene	50/1/0.05	3	>99	5.32	5.63	1.20
10[h]	(*Z*)-DATP	Toluene	50/1/0.05	4	>99	10.5	11.0	1.14

[a]$[M]_0$ = 0.50 mol L^{-1}. [b]Determined by ^1H NMR in $CDCl_3$. [c]Calculated from the equation described in Note 1. [d]Determined by SEC in DMF containing 0.01 mol L^{-1} LiBr using a multi-angle laser light scattering detector. [e]Determined by SEC in DMF containing 0.01 mol L^{-1} LiBr using polystyrene standards with an RI detector. [f]$[M]_0$ = 0.80 mol L^{-1}. [g]$[M]_0$ = 2.0 mol L^{-1}. [h]Post polymerization after run 9

3.3 Results and Discussion

3.3.1 *Initiator for Tf_2NH-Promoted GTP of DMAA*

First, the author carried out the polymerization of DMAA using 1-methoxy-1-trimethylsiloxy-2-methyl-1-propene (MTS), the conventional GTP initiator, in CH_2Cl_2 and toluene (Table 3.1, runs 1 and 2). Unexpectedly, the number-average molecular weights (M_n) of the obtained polymers were 20.2 and 23.7 kg mol^{-1}, which were extremely higher than those previously prepared by the GTP method [1, 11–14]. In particular, for the polymerization in CH_2Cl_2, the monomer was quantitatively consumed after 2 h and the molecular weight distribution (M_w/M_n) value of the obtained polymer was relatively low as 1.23, though the M_n of the polymer was twice of the predicted value ($M_{n,calcd.}$) calculated from the monomer conversion and the initial concentration of the monomer, initiator, and Tf_2NH ($[M]_0$, $[I]_0$, and $[Tf_2NH]_0$).[1] These results obviously indicated that Tf_2NH was an effective catalyst for the polymerization of DMAA even using MTS as the initiator.

[1] The theoretical molecular weight ($M_{n,calcd}$) of PDMAA obtained from Tf_2NH-promoted GTP was calculated by the following equation: $M_{n,calcd} = [M]_0/([I]_0-[Tf_2NH]_0) \times \text{conv.} \times$ (MW of DMAA = 99.13) + (MW of initiator residue = 101.15). According to the proposed mechanism of polymerization described in Ref. [25] the silyl enolate for the initiator must be consumed by the reaction with Tf_2NH to produce the actual catalyst, Tf_2NSiMe_3. Thus, $[Tf_2NH]_0$ was subtracted from $[I]_0$ in the equation to obtain the effective concentration of the initiator for the polymerization

Table 3.2 GTP of DMAA using (*Z*)-DATP and conventional catalysts or Tf_2NH in toluene at 0 °C for 6 h[a] (Adapted with permission from Ref. [26]. Copyright 2010 American Chemical Society)

Run	Catalyst	Conv. (%)[b]	M_n (kg mol^{-1})		M_w/M_n^e
			Calcd.[c]	Obsd.[d]	
11	TAS-HF_2	0	–	–	–
12	TBA-AcO	57.0	5.75	340	1.63
13	ZnI_2	0	–	–	–
14	Tf_2NH	>99	10.2[f]	12.1	1.09

[a] $[M]_0 = 0.50$ mol L^{-1}; $[M]_0/[I]_0/[cat.]_0 = 100/1/0.02$. [b] Determined by ^{1}H NMR in $CDCl_3$. [c] Calculated from $[M]_0/[I]_0 \times$ conv. $\times$ (MW of DMAA) + (MW of (*Z*)-DATP). [d] Determined by SEC in DMF containing 0.01 mol L^{-1} LiBr using a multiangle laser light scattering detector. [e] Determined by SEC in DMF containing 0.01 mol L^{-1} LiBr using polystyrene standards with an RI detector. [f] Calculated from the equation described in Note 1

Thus, the author designed and synthesized (*Z*)-1-dimethylamino-1-trimethylsiloxy-1-propene ((*Z*)-DATP) as the new initiator for the GTP of acrylamide monomers. In principle, the rate of initiation should be greater or equal to the rate of propagation to realize living polymerization. As expected from the structural homology of the initiating species and growing species, the polymerization using (*Z*)-DATP smoothly proceeded in CH_2Cl_2 and toluene to afford well-defined PDMAAs with the low polydispersities of 1.08 and 1.11 (Table 3.1, runs 3 and 4). In addition, the M_ns were well-controlled as 12.7 kg mol^{-1} for CH_2Cl_2 and 11.9 kg mol^{-1} for toluene, which agreed with the $M_{n,calcd.}$ of 10.2 and 10.5 kg mol^{-1}, respectively. These results strongly suggested that the structure of the initiator significantly affects the initiation character of the Tf_2NH-promoted GTP of DMAA.

In order to directly prove that the combination of (*Z*)-DATP and Tf_2NH is exceptionally effective for the GTP of DMAA, the author next evaluated the combinations of (*Z*)-DATP and conventional catalysts, such as zinc iodide (ZnI_2) as the Lewis acid and tris(dimethylamino)sulfonium bifluoride (TAS-HF_2) and tetra-*n*-butylammonium acetate (TBA-AcO) as the Lewis bases. All the polymerizations were carried out with $[DMAA]_0/[(Z)\text{-DATP}]_0/[catalyst]_0 = 100/1/0.02$ in toluene at 0 °C for 6 h. Table 3.2 summarizes the polymerization results. In clear contrast to the GTP of MMA [11], the Lewis basic catalyst of TAS-HF_2 showed almost no catalytic activity toward the GTP of DMAA (conv. ~0 %, Table 3.2, run 11). Although TBA-AcO, another Lewis basic catalyst, showed a fairly high monomer conversion of 57.0 % (Table 3.2, run 12), the M_n of the obtained PDMAA (340 kg mol^{-1}) was extremely higher than that of $M_{n,calcd.}$ (5.75 kg mol^{-1}) and the M_w/M_n was as broad as 1.63, showing that conventional Lewis basic catalysts were unsuitable for the GTP of DMAA. ZnI_2, an active catalyst for the GTP of acrylates, also showed almost no catalytic activity (conv. ~0 %, Table 3.2, run 13). In clear contrast to these conventional catalysts, DMAA was quantitatively consumed in the Tf_2NH-promoted GTP, obviously indicating that the new GTP initiating system consisting of (*Z*)-DATP and Tf_2NH was exceptionally effective for the GTP of DMAA as the representative acrylamide monomer.

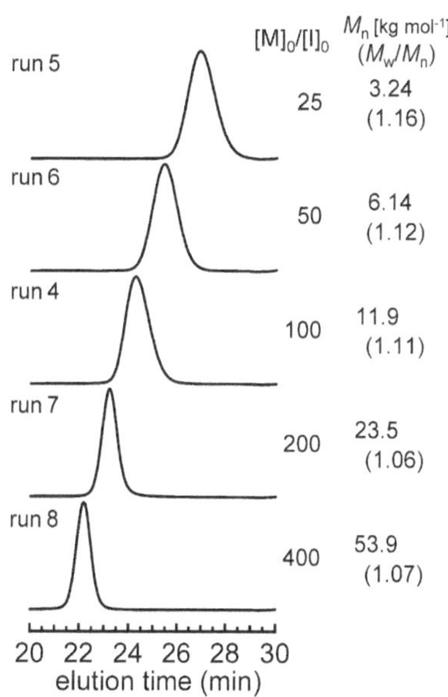

Fig. 3.2 SEC traces of the PDMAA polymerized with different monomer to initiator ratios ($[M]_0/[I]_0$) (runs 4–8) (eluent, DMF containing 0.01 mol L^{-1} LiBr; flow rate, 1.0 mL min^{-1}). (Adapted with permission from Ref. [26]. Copyright 2010 American Chemical Society)

3.3.2 *Living Characteristics of Tf_2NH-Promoted GTP of DMAA*

In order to characterize the living nature of the GTP of DMAA, the author first carried out the polymerizations using various ratios of $[DMAA]_0/[(Z)\text{-}DATP]_0$ from 25 to 400 (Table 3.1, runs 4–8). Figure 3.2 shows the SEC traces of the obtained PDMAAs. Their M_ns increased from 3.24 to 53.9 kg mol^{-1}, which agreed with the $M_{n,calcd}$ values predicted from the $[DMAA]_0/[(Z)\text{-}DATP]_0$. Furthermore, all the M_w/M_n values were as low as 1.06–1.16, which strongly indicated that the Tf_2NH-promoted GTP of DMAA proceeded in a living manner. In addition, the living nature was confirmed from the kinetic study. As shown in Fig. 3.3a, the molecular weight of the obtained PDMAA linearly increased from 2.82 to 8.88 kg mol^{-1} with the increasing monomer conversion while the M_w/M_n values of the obtained PDMAAs retained low values in the range of 1.09–1.16. The first-order kinetic plots shown in Fig. 3.3b proved the distinct first-order relationship between the reaction time and the monomer conversion. In order to provide a detailed insight into the polymerization reaction, a MALDI-TOF MS measurement for the obtained PDMAA was carried out. The results were summarized in Fig. 3.4. Only one series of peaks was observed as shown in Fig. 3.4a, b. The difference in the *m/z* values among each molecular ion peak was just 99.1, which corresponded to the molecular weight of DMAA as a monomer unit. An expanded molecular ion peak in the obtained MS spectrum was shown in Fig. 3.4c. The pattern of the observed isotopic distribution of the molecular ion peak was in a good agreement with the theoretical isotopic distribution of PDMAA consisting of 39 repeating units, a residue of (*Z*)-DATP, and

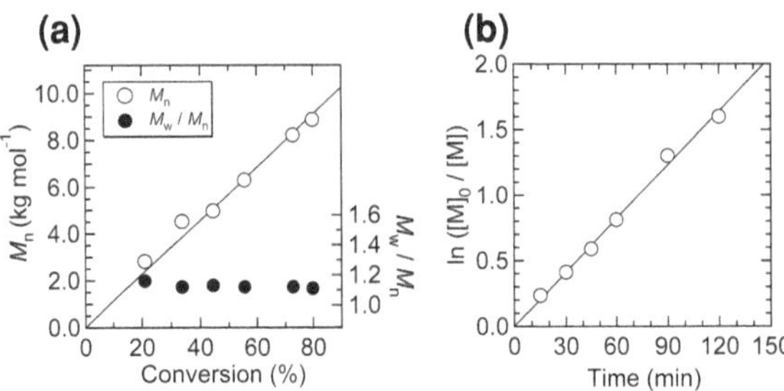

Fig. 3.3 (**a**) Dependence of molecular weight (M_n) and polydispersity index (M_w/M_n) on the monomer conversion, and (**b**) kinetic plots for the polymerization of DMAA ([DMAA]$_0$ = 0.50 mol L^{-1}; [DMAA]$_0$/[(*Z*)-DATP]$_0$/[Tf$_2$NH]$_0$ = 100/1/0.02; solvent, toluene; temperature, 0 °C). (Adapted with permission from Ref. [26]. Copyright 2010 American Chemical Society)

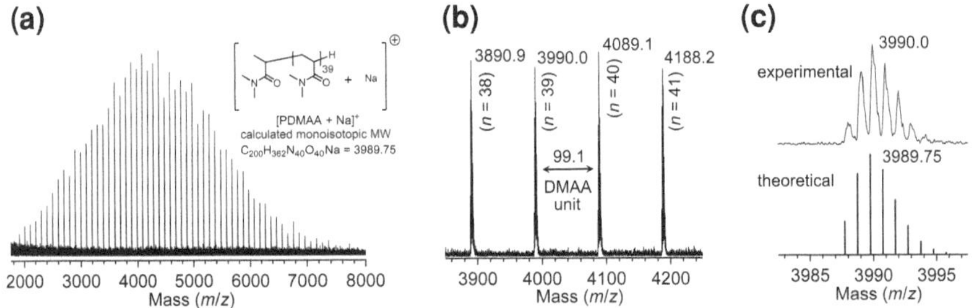

Fig. 3.4 MALDI-TOF MS spectra measured in reflector mode of the obtained PDMAA ([M]$_0$/[I]$_0$/[Tf$_2$NH]$_0$ = 100/1/0.02, conversion = 44.5 %, M_n = 4.97 kg mol^{-1}, M_w/M_n = 1.13). (Adapted with permission from Ref. [26]. Copyright 2010 American Chemical Society)

the desilylated propagating end cationized by sodium ions (molecular formula: $C_{200}H_{362}O_{40}N_{40}Na$). These results indicated that Tf$_2$NH-promoted GTP proceeded in a living fashion without any side reactions, such as a back biting reaction [19, 28]. Finally, the chain extension experiment also proved the living nature of the Tf$_2$NH-promoted GTP of DMAA. Figure 3.5 shows SEC traces for the chain extension experiment. A PDMAA with M_n = 5.63 kg mol^{-1} and M_w/M_n = 1.20 was first prepared by the perfect conversion of 50 equivalents of DMAA under the conditions of [(*Z*)-DATP]$_0$/[Tf$_2$NH]$_0$ = 1/0.05 in toluene at 0 °C. The polymerization was further continued by adding 50 equivalents of DMAA to afford a PDMAA with M_n = 11.1 kg mol^{-1} and M_w/M_n = 1.14 (run 10), indicating that the propagating end of the growing PDMAA possessed a truly living nature. Thus, the author for the first time achieved the living polymerization of acrylamide through the GTP process by using the strong Brønsted acid, Tf$_2$NH, as a good promoter of the polymerization.

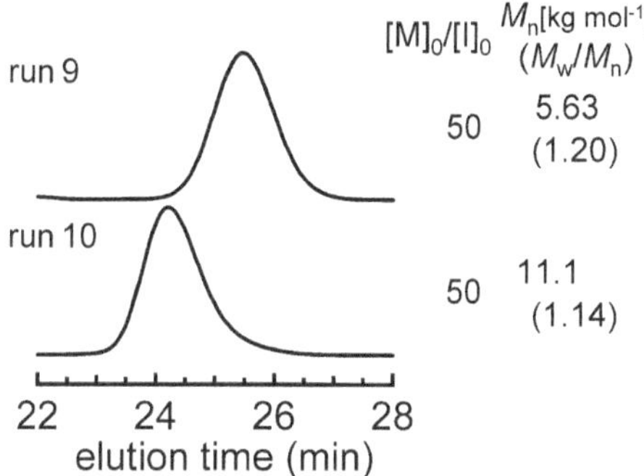

Fig. 3.5 SEC traces of the PDMAA obtained by the first polymerization (run 9) and following post-polymerization (run 10) (eluent, DMF containing 0.01 mol L^{-1} of LiBr; flow rate, 1.0 mL min^{-1}). (Adapted with permission from Ref. [26]. Copyright 2010 American Chemical Society)

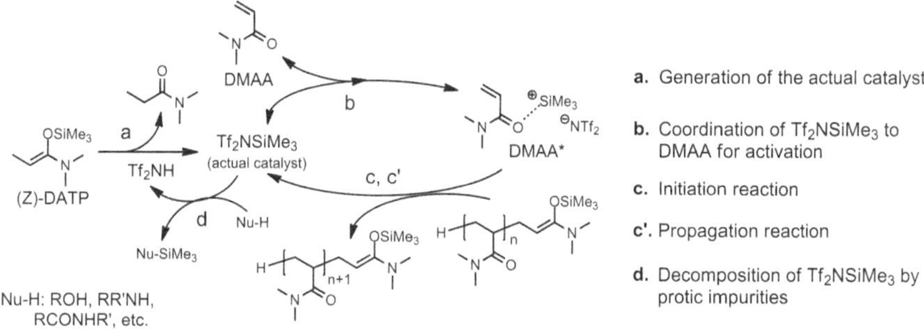

Scheme 3.2 Presumed mechanism of Tf_2NH-promoted GTP of DMAA

3.3.3 Mechanism of the Tf_2NH-Promoted GTP of DMAA

The mechanism of the Tf_2NH-promoted GTP of DMAA GTP was presumed based on the strong Brønsted acid-promoted Mukaiyama aldol reaction [29–31], as shown in Scheme 3.2. Five elementary reactions were defined: (a) the generation of *N*-(trimethylsilyl)bis(trifluoromethanesulfonyl)imide (Tf_2NSiMe_3) as the actual catalyst from the irreversible reaction between Tf_2NH and (*Z*)-DATP; (b) the reversible coordination of Tf_2SiMe_3 to DMAA for the formation of an activated monomer (DMAA*); (c and d) the initiation/propagation reaction between DMAA* and the silyl enolate in (*Z*)-DATP or the propagating polymer; and (e) the decomposition of Tf_2NSiMe_3 with protic impurities (Nu-H) accompanied by the regeneration of Tf_2NH. Thus, the polymerization was considered to proceed through the activated monomer mechanism as with the Lewis acid catalyzed-GTP [32].

In order to confirm the occurrence of reaction b, the ^1H NMR spectra of DMAA, Tf_2NSiMe_3, as well as an equimolar mixture of DMAA and Tf_2NSiMe_3 were measured in $CDCl_3$ as shown in Fig. 3.6 ([DMAA] = [Tf_2NSiMe_3] = 0.16

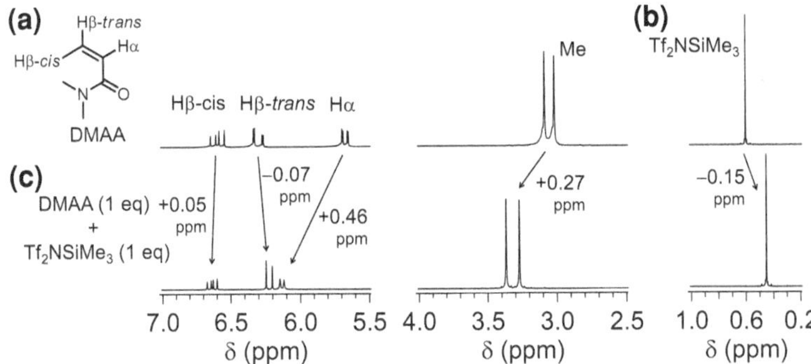

Fig. 3.6 ^{1}H NMR spectra of **(a)** DMAA, **(b)** Tf_2NSiMe_3, and **(c)** equimolar mixture of DMAA and Tf_2NSiMe_3, in $CDCl_3$ at room temperature. ([DMAA] = [Tf_2NSiMe_3] = 0.167 mol L^{-1})

7 mol L^{-1}). The resonance due to the α-proton, the methyl group, and the β-*cis* proton of the DMAA underwent 0.40, 0.27, and 0.05 ppm low-field shifts after mixing with Tf_2NSiMe_3. The β-*trans* proton showed a 0.07 ppm upfield shift. The resonance due to the trimethylsilyl group of Tf_2NSiMe_3 moved from 0.60 to 0.44 ppm and had a 0.16 ppm upfield shift after mixing with DMAA. Moreover, no new peaks emerged after mixing DMAA and Tf_2NSiMe_3 Thus, these results proved that the coordination of Tf_2NSiMe_3 to DMAA actually occurs [33].

3.3.4 Stereoregularity and Thermal Properties of PDMAA

The stereoregularity of a polymer, one of the important structural factors, affects the properties of a polymer, such as the melting temperature, glass transition temperature, and mechanical strength. It is interesting to compare the stereoregularity of PDMAAs synthesized by the Tf_2NH-promoted GTP with those prepared using other polymerization methods because the activated monomer mechanism of the Tf_2NH-promoted GTP has the possibility to induce a high stereoregularity to the resultant PDMAA. Figure 3.7 shows the ^{1}H and ^{13}C NMR spectra of PDMAAs synthesized using (*Z*)-DATP as the initiator and Tf_2NH as the catalyst. The diad contents (*m*/*r*) of PDMAA were determined from the resonances of the main-chain methylene protons appearing between 1 and 2 ppm in the ^{1}H NMR spectra measured in DMSO-d_6 at 150 °C [4–6, 34, 35]. The *m*/*r* values of the PDMAAs polymerized at 0 °C in CH_2Cl_2 and toluene (Table 3.1, runs 3 and 4) were 35/65 and 44/56, respectively, and those polymerized at −78 °C in CH_2Cl_2 and toluene were 27/73 and 38/62, respectively. The triad contents (*mm*/*mr*/*rr*) of PDMAA were analyzed from the carbonyl carbon appearing around 175 ppm in the ^{13}C NMR spectra measured in $CDCl_3$ at 50 °C. The contents of the *mm* and *mr* triad decreased and that of the *rr* triad increased when the polymerization was carried out in CH_2Cl_2 or at a low temperature. These results indicated that the Tf_2NH-promoted GTP of DMAA in a polar solvent or at a lower temperature produced

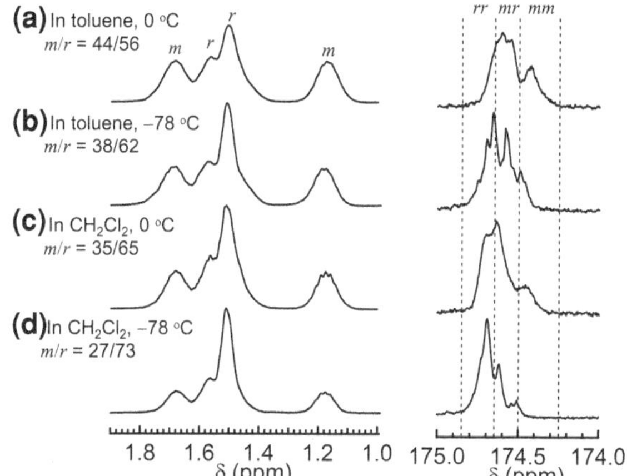

Fig. 3.7 ^{1}H NMR spectra measured in DMSO-d_6 at 150 °C (*left*) and ^{13}C NMR spectra measured in $CDCl_3$ at 50 °C (*right*) of the PDMAAs obtained in (**a**) toluene at 0 °C, (**b**) in toluene at −78 °C, (**c**) in CH_2Cl_2 at 0 °C, and (**d**) in CH_2Cl_2 at −78 °C. (Adapted with permission from Ref. [26]. Copyright 2010 American Chemical Society)

the *r* diad rich PDMAA. The dependence of the tacticity on the polymerization temperature was different from that for the conventional radical polymerization, in which the content of the *m* diad of the resultant PDMAA increased with the decreasing polymerization temperature [35, 36].

Despite the importance of the glass transition temperature (T_g) in polymer science, detailed characterization of the T_g of PDMAA has been scarcely reported [37–41]. The author takes into account that PDMAA synthesized by Tf_2NH-promoted GTP of DMAA was able to produce an ideal model polymer; in other words, PDMAA synthesized by Tf_2NH-promoted GTP of DMAA possessed no initiator and terminator residues. Hence, the author finally focused on the characterization of the T_g for PDMAAs with various molecular weights. Figure 3.8 shows a typical DSC trace for PDMAA. The step in the baseline is assignable to a glass transition. The glass transition temperature was defined as the inflection point in the step and was determined from the minimum of the first derivative of the curve. Basically, the T_g of PDMAA has been known to be around 120 °C [38, 39]. Fig. 3.9 shows the plots of the T_gs as a function of the M_n, indicating that the T_gs increased with the increasing M_n values. Importantly, the T_g for the PDMAA possessed a molecular weight dependence below 20 kg mol^{-1}. Thus, precise characterization of the glass transition temperature for PDMAA was successfully achieved using Tf_2NH-promoted GTP of DMAA.

3.4 Conclusions

Tf_2NH, a strong Brønsted acid, was revealed to promote the GTP of DMAA. The initiating system comprised of (*Z*)-DATP and Tf_2NH was able to produce a PDMAA with a higher molecular weight than that of any previously synthesized polyacrylamides by the GTP using other initiating systems. As a direct consequence of the homology in the structure of an initiator and the chain end of the

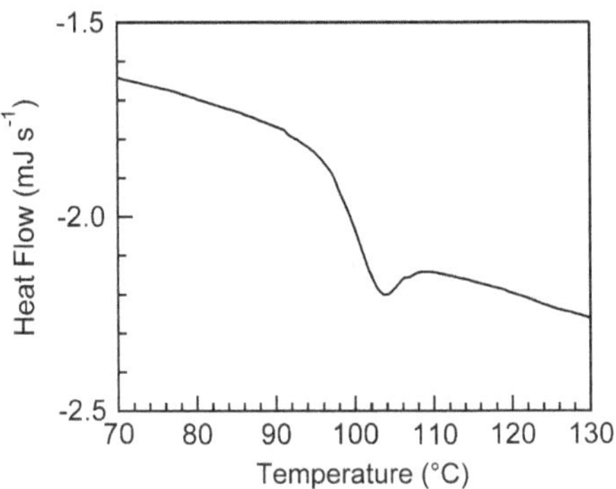

Fig. 3.8 Differential scanning calorimetry (DSC) trace for PDMAA ($M_n = 4.52$ kg mol^{-1}, $M_w/M_n = 1.12$). (Adapted with permission from Ref. [26]. Copyright 2010 American Chemical Society)

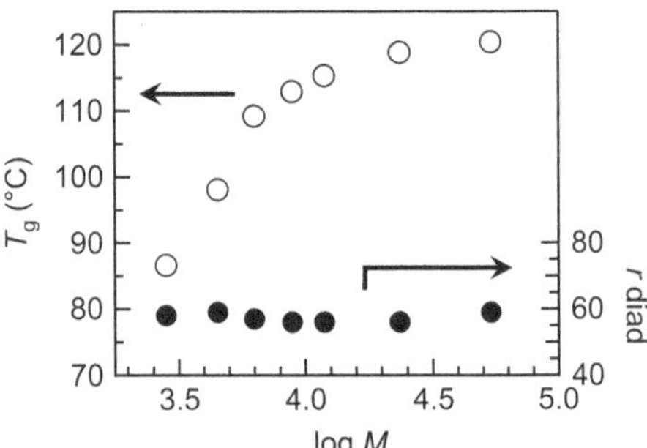

Fig. 3.9 The plots of the glass transition temperature (T_g) and *r* diad content for PDMAA as a function of the number averaged-molecular weight (M_n). (Adapted with permission from Ref. [26]. Copyright 2010 American Chemical Society)

propagating polymer (*Z*)-DATP, an amino silyl enolate, showed an excellent initiating efficiency and polymerization control as the GTP initiator in comparison to the conventional initiator of MTS. In addition, the living character of the GTP of DMAA using (*Z*)-DATP and Tf_2NH was strongly confirmed by the kinetic study, the MALDI-TOF MS measurement, and the post-polymerization experiment. To the best of the author's knowledge, this study was the first demonstration of the living polymerization of acrylamide monomers by the GTP. This newly developed GTP was considered to be valuable for the precise synthesis of thermoresponsive polyacrylamides because the polymerization produced a polymer with a predetermined molecular weight, an extremely narrow polydispersity, and a well-defined chain-end structure.

References

1. Webster OW, Hertler WR, Sogah DY, Farnham WB, RajanBabu TV (1983) Group-transfer polymerization. 1. A new concept for addition polymerization with organosilicon initiators. J Am Chem Soc 105:5706–5708
2. Webster OW (2000) The discovery and commercialization of group transfer polymerization. J Polym Sci, Part A: Polym Chem 38:2855–2860
3. Webster OW (2004) Group transfer polymerization: Mechanism and comparison with other methods for controlled polymerization of acrylic monomers. Adv Polym Sci 167:1–34

4. Kobayashi M, Okuyama S, Ishizone T, Nakahama S (1999) Stereospecific anionic polymerization of *N*,*N*-dialkylacrylamides. Macromolecules 32:6466–6477
5. Kobayashi M, Ishizone T, Nakahama S (2000) Synthesis of highly isotactic poly(*N*,*N*-diethylacrylamide) by anionic polymerization with grignard reagents and diethylzinc. J Polym Sci, Part A: Polym Chem 38:4677–4685
6. Kobayashi M, Ishizone T, Nakahama S (2000) Additive effect of triethylborane on anionic polymerization of *N*,*N*-dimethylacrylamide and *N*,*N*-diethylacrylamide. Macromolecules 33:4411–4416
7. Ishizone T, Ito M (2002) Synthesis of well-defined Poly(*N*-isopropylacrylamide) by the anionic polymerization of *N*-methoxymethyl-*N*-isopropylacrylamide. J Polym Sci, Part A: Polym Chem 40:4328–4332
8. Ito M, Ishizone T (2006) Living anionic polymerization of *N*-methoxymethyl-*N*-isopropylacrylamide: Synthesis of well-defined poly(*N*-isopropylacrylamide) Having Various Stereoregularity. J Polym Sci, Part A: Polym Chem 44:4832–4845
9. Suzuki T, Kusakabe J, Ishizone T (2008) Living anionic polymerization of *N*-methacryloyl-2-methylaziridine: Polymerizable *N*,*N*-dialkylmethacrylamide. Macromolecules 41:1929–1936
10. Suzuki T, Kusakabe J-I, Kitazawa K, Nakagawa T, Kawauchi S, Ishizone T (2010) Living anionic polymerization of *N*-methacryloylazetidine: Anionic polymerizability of *N*,*N*-dialkylmethacrylamides. Macromolecules 43:107–116
11. Sogah DY, Hertler WR, Webster OW, Cohen GM (1987) Group transfer polymerization - polymerization of acrylic monomers. Macromolecules 20:1473–1488
12. Eggert M, Freitag R (1994) Poly-*N*,*N*-diethylacrylamide prepared by group transfer polymerization: synthesis, characterization, and solution properties. J Polym Sci, Part A: Polym Chem 32:803–813
13. Freitag R, Baltes T, Eggert M (1994) A comparison of thermoreactive water-soluble poly-*N*,*N*-diethylacrylamide Prepared by anionic and by group transfer polymerization. J Polym Sci, Part A: Polym Chem 32:3019–3030
14. Baltes T, Garret-Flaudy F, Freitag R (1999) Investigation of the LCST of polyacrylamides as a function of molecular parameters and the solvent composition. J Polym Sci, Part A: Polym Chem 37:2977–2989
15. Culkin DA, Jeong WH, Csihony S, Gomez ED, Balsara NR, Hedrick JL, Waymouth RM (2007) Zwitterionic polymerization of lactide to cyclic poly(lactide) by using *N*-heterocyclic carbene organocatalysts. Angew Chem Int Ed 46:2627–2630
16. Raynaud J, Absalon C, Gnanou Y, Taton D (2009) *N*-heterocyclic carbene-induced zwitterionic ring-opening polymerization of ethylene oxide and direct synthesis of alpha,omega-difunctionalized poly(ethylene oxide)s and poly(ethylene oxide)-*b*-poly(epsilon-caprolactone) block copolymers. J Am Chem Soc 131:3201–3209
17. Raynaud J, Ciolino A, Baceiredo A, Destarac M, Bonnette F, Kato T, Gnanou Y, Taton D (2008) Harnessing the potential of *N*-heterocyclic carbenes for the rejuvenation of group-transfer polymerization of (meth) acrylics. Angew Chem Int Ed 47:5390–5393
18. Scholten MD, Hedrick JL, Waymouth RM (2008) Group transfer polymerization of acrylates catalyzed by *N*-heterocyclic carbenes. Macromolecules 41:7399–7404
19. Raynaud J, Gnanou Y, Taton D (2009) Group transfer polymerization of (meth)acrylic monomers catalyzed by *N*-heterocyclic carbenes and synthesis of all acrylic block copolymers: Evidence for an associative mechanism. Macromolecules 42:5996–6005
20. Kakuchi R, Chiba K, Fuchise K, Sakai R, Satoh T, Kakuchi T (2009) Strong brønsted acid as a highly efficient promoter for group transfer polymerization of methyl methacrylate. Macromolecules 42:8747–8750
21. Foropoulos J, Desmarteau DD (1984) Synthesis, properties, and reactions of bis((trifluoromethyl)sulfonyl)imide, $(CF_3SO_2)_2NH$. Inorg Chem 23:3720–3723
22. Ishihara K, Hiraiwa Y, Yamamoto H (2001) A high yield procedure for the Me_3SiNTf_2-induced carbon-carbon bond-forming reactions of silyl nucleophiles with carbonyl compounds: The Importance of addition order and solvent effects. Synlett 1851–1854
23. Hiraiwa Y, Ishihara K, Yamamoto H (2006) Crucial role of the conjugate base for silyl Lewis ccid Induced Mukaiyama aldol reactions. Eur J Org Chem 2006:1837–1844

24. Akiyama T (2007) Stronger brønsted acids. Chem Rev 107:5744–5758
25. Yamamoto H (2007) New reaction and new catalyst—a personal perspective. Tetrahedron 63:8377–8412
26. Fuchise K, Sakai R, Satoh T, Sato S, Narumi A, Kawaguchi S, Kakuchi T (2010) Group transfer polymerization of *N*,*N*-dimethylacrylamide using novel efficient system consisting of dialkylamino silyl enol ether as an Initiator and strong brønsted acid as an organocatalyst. Macromolecules 43:5589–5594
27. Kikuchi M, Lien LTN, Narumi A, Jinbo Y, Izumi Y, Nagai K, Kawaguchi S (2008) Conformational properties of cylindrical rod brushes consisting of a polystyrene main chain and poly(*n*-hexyl isocyanate) side chains. Macromolecules 41:6564–6572
28. Müller AHE (1990) Group transfer and anionic polymerization: A critical comparison. Makromol Chem, Macromol Symp 32:87–104
29. Ishihara K, Hiraiwa Y, Yamamoto H (2002) Crucial role of the ligand of silyl Lewis acid in the Mukaiyama aldol reaction. Chem Commun 1564–1566
30. Hiraiwa Y, Ishihara K, Yamamoto H (2006) Crucial role of the conjugate base for silyl Lewis acid Induced Mukaiyama aldol reactions. Eur J Org Chem 2006:1837–1844
31. García-García P, Lay F, García-García P, Rabalakos C, List B (2009) A powerful chiral counteranion motif for asymmetric catalysis. Angew Chem Int Ed 48:4363–4366
32. Hertler WR, Sogah DY, Webster OW (1984) Group-transfer polymerization. 3. Lewis acid catalysis. Macromolecules 17:1415–1417
33. Hasegawa A, Ishihara K, Yamamoto H (2003) Trimethylsilyl pentafluorophenylbis(trifluorom ethanesulfonyl) methide as a super Lewis acid catalyst for the condensation of trimethylhydroquinone with Isophytol. Angew Chem Int Ed 42:5731–5733
34. Bulai A, Jimeno ML, De Queiroz AAA, Gallardo A, San Román J (1996) ^{1}H and ^{13}C nuclear magnetic resonance studies on the stereochemical configuration of bis(*N*,*N*-dimethyl-2,4-dimethylglutarylamide) and Poly(*N*,*N*-dimethylacrylamide). Macromolecules 29:3240–3246
35. Liu WH, Nakano T, Okamoto Y (2000) Stereocontrol in radical polymerization of *N*,*N*-dimethylacrylamide and *N*,*N*-diphenylacrylamide and thermal properties of syndiotactic poly(methyl acrylate)s serived from the obtained polymers. Polym J 32:771–777
36. Hirano T, Masuda S, Nasu S, Ute K, Sato T (2009) Syndiotactic-specific radical polymerization of *N*,*N*-dimethylacrylamide in the presence of tartrates: A proposed mechanism for the polymerization. J Polym Sci, Part A: Polym Chem 47:1192–1203
37. Krause S, Gormley JJ, Roman N, Shetter JA, Watanabe WH (1965) Glass temperatures of some acrylic polymers. J Polym Sci Part A Gen Pap 3:3573–3586
38. Mohajer Y, Wilkes GL, Gia H, McGrath JE (1980) Influence of molecular geometry on the mechanical properties of homopolymers and block polymers of hydrogenated butadiene and Isoprene. Polym Prep (Am Chem Soc Div Polym Chem) 21:229–230
39. Xie X, Hogen-Esch TE (1996) Anionic synthesis of narrow molecular weight distribution water-soluble poly(*N*,*N*-dimethylacrylamide) and poly(*N*-acryloyl-*N*′-methylpiperazine). Macromolecules 29:1746–1752
40. Bolig AD, Chen EY-X (2004) *ansa*-Zirconocene ester enolates: synthesis, structure, reaction with organo-Lewis acids, and application to polymerization of methacrylates. J Am Chem Soc 126:4897–4906
41. Mariott WR, Chen EY-X (2005) Mechanism and scope of stereospecific, coordinative-Anionic polymerization of acrylamides by chiral zirconocenium ester and amide enolates. Macromolecules 38:6822–6832

Chapter 4
Facile Synthesis of Thermoresponsive Block Copolymers Bearing Poly (*N*,*N*-diethylacrylamide) Segment Through Group Transfer Polymerization

4.1 Introduction

Living polymerizations have enabled the precise synthesis of well-defined polymers including end-functionalized polymers, block copolymers, star-shaped polymers, macrocyclic polymers, etc., which have allowed controlling the properties of the polymers through designing the polymer structure. The precise synthesis of not only thermoresponsive polymers, but also their block copolymers, namely the thermoresponsive amphiphilic block copolymers and double-hydrophilic block copolymers, has also been achieved through the development of controlled/living polymerizations, as described in Chap. 1 [1–7]. The following synthetic strategies have been applied to the precise synthesis of thermoresponsive block copolymers including those bearing one or multiple polyacrylamide segments: (1) sequential polymerization of multiple monomers (2) a polymerization using a macroinitiator or a macro chain transfer agent (macro-CTA), and (3) the polymer-polymer conjugation reaction. The first approach requires that the polymerization can be continued until the monomer is quantitatively consumed and the propagating end retains reactivity even after completion of the polymerization. Thus, only the anionic polymerization has been suitable for the synthesis of block copolymers bearing thermoreponsive polyacrylamides by the first approach. For example, Ishizone et al. reported the sequential polymerization of styrene or isoprene and *N*-isopropyl-*N*-methoxymethylacrylamide, namely the *N*-isopropylacrylamide (NIPAM) protected by a methoxymethyl group, to synthesize the block copolymers bearing poly(*N*-isopropylacrylamide) (PNIPAM) segments [8]. Müller et al. [9, 10] Adler et al. [11] and Tsitsilianis et al. [12] independently reported the sequential anionic polymerization of *tert*-butyl acrylate (*t*BA), *tert*-butyl methacrylate (*t*BMA), and 2-vinylpyridine (2VP) as the first monomer and *N*,*N*-diethylacrylamide (DEAA) as the second monomer. The second approach is widely applied to the synthesis of block copolymers by controlled/living radical polymerizations, such as the nitroxide-mediated polymerization (NMP) [13, 14], atom transfer radical polymerization (ATRP) [15–19], and reversible addition-fragmentation chain transfer (RAFT) polymerization [4, 20–31], and a ring-opening polymerization

K. Fuchise, *Design and Precise Synthesis of Thermoresponsive Polyacrylamides*, Springer Theses, DOI: 10.1007/978-4-431-55046-4_4,

(ROP) [32] using a macroinitiator or a macro-CTA. Various click reactions have been utilized for the third approach; for example, Thayumanavan et al. reported the synthesis of an amphiphilic thermoresponsive block copolymer possessing a PNIPAM segment by the pyridyl disulfide exchange reaction [33].

The author has discovered that the GTP of *N*,*N*-dimethylacrylamide (DMAA) using trifluoromethanesulfoimide (Tf_2NH) and an amino silyl enolate proceeded in a living fashion to produce poly(*N*,*N*-dimethylacrylamide) (PDMAA) with a high molecular weight and a narrow polydispersity, as described in Chap. 3 [34]. This polymerization was expected to become a robust method to precisely synthesize other poly(*N*,*N*-dialkylacrylamide)s with thermoresponsive properties and their block copolymers. Poly(*N*,*N*-diethylacrylamide) (PDEAA) is one of the representative thermoreponsive polyacrylamides, and its lower critical solution temperature (LCST) is around 32 °C as with PNIPAM. Several groups have reported the synthesis of amphiphilic thermoresponsive block copolymers and double-hydrophilic block copolymers bearing a PDEAA segment by the sequential anionic polymerization [9–12] and the RAFT polymerization using a macro-CTA [23–27, 29, 30, 32]. The Tf_2NH-promoted GTP has shown that the polymerization can be continued until the monomer is quantitatively consumed unlike the controlled/living radical polymerizations, which was considered to allow synthesizing a block copolymer just by the sequential polymerization as with the anionic polymerization. The synthesis of a block copolymer using a macroinitiator was also considered to be achievable through the synthesis of end-functionalized polyacrylamides by the GTP using a functional amino silyl enolate. The author focused on the block copolymers of PDEAA and the polyethers derived from epoxides because the polyethers have different solubilities in water depending on the side-chain structure and the molecular weight; for example, poly(ethylene oxide) (PEO) is water-soluble, poly(propylene oxide) (PPO) varies from water-soluble to water-insoluble as the molecular weight increases, and poly(1,2-butene oxide) (PBO) is water-insoluble. The controlled/living polymerization of various epoxides has been developed using an alcohol as the initiator and a phosphazene base, 1-*tert*-butyl-4,4,4-tris(dimethylamino)-2,2-bis-[tris(dimethylamino)-phosphoranylidenamino]-$2\Lambda^5$,$4\Lambda^5$-catenadi(phosphazene) (*t*-Bu-P_4), as the catalyst [35–40]. Furthermore, the synthesis of the block copolymers having a polyether segment has been achieved by the *t*-BuP_4-catalyzed ROP of ethylene oxide (EO) using hydroxyl end-functionalized poly(butadiene) [37] and poly(isobutene) [38]. Thus, the precise synthesis of PDEAA-*b*-polyether was expected to be achieved by the *t*-Bu-P_4-catalyzed ROP of epoxides using the hydroxyl end-functionalized PDEAA (PDEAA-OH), which was considered to provide a robust and facile method for the synthesis of thermoresponsive block copolymers with various properties.

In this study, the author aimed to develop a versatile method to prepare amphiphilic thermoresponsive polymers and double-hydrophilic polymers possessing a PDEAA segment through the sequential GTP of two acrylamide monomers as well as the synthesis of the macroinitiator by the GTP. First, the author describes the precise synthesis of PDEAA and PDEAA-OH by the Tf_2NH-promoted GTP using amino silyl enolates, namely (*Z*)-1-dimethylamino-1-trimethylsiloxy-1-propene ((*Z*)-DATP) and (*Z*)-1-(*N*-

methyl-*N*-(2-trimethylsiloxyethyl)amino)-1-trimethylsiloxy-1-propene ((*Z*)-HDATP). The synthesis of PDEAA-*b*-PDMAA was achieved by the sequential polymerization of DEAA and DMAA. The synthesis of the PDEAA-*block*-polyethers was achieved by the *t*-Bu-P_4-catalyzed ROP of epoxides using PDEAA-OH as the macroinitiator. The thermoresponsive properties of the resultant PDEAA-*block*-polyethers in an aqueous solution were investigated to clarify the effect of the introduced polyether segments on the thermoresponsive properties of PDEAA-*block*-polyether.

4.2 Experimental Section

4.2.1 Materials

2-(Methylamino)ethanol (97.0 %), bis(trifluoromethanesulfonyl)imide (Tf_2NH, >95 %), dichloromethane (CH_2Cl_2, >99.5 %; water content, <0.001 %), toluene (>99.5 %; water content, <0.001 %), tetrahydrofuran (THF, >99.5 %; water content, <0.001 %), propionyl chloride, *n*-butyllithium (1.6 mol L^{-1} in *n*-hexane), *N*,*N*,*N*′,*N*′-tetramethylethylenediamine (TMEDA), potassium carbonate (K_2CO_3), methanol, 2-propanol (>99.5 %), *tert*-butyl alcohol (*t*BuOH) (>98.0 %), acetonitrile (>99.5 %), pyridine (>99.0 %), and dimethylsulfoxide-d_6 (DMSO-d_6) were purchased from Kanto Chemicals Co., Inc. 1,2-Butene oxide (BO), ethylene oxide (EO, 1.2 mol L^{-1} in THF), *N*,*N*-diethylacrylamide (DEAA), *N*,*N*-dimethylacrylamide (DMAA), 1-methylimidazole, diisopropylamine, chlorotrimethylsilane, and *trans*-3-indoleacrylic acid were purchased from Tokyo Kasei Kogyo Co., Ltd. 1-*tert*-Butyl-4,4,4-tris(dimethylamino)-2,2-bis[tris(dimethylamino)phosphoranylidenamino]-2Λ^5,4Λ^5-catenadi(phosphazene) (*t*-Bu-P_4) (1.0 mol L^{-1} in *n*-hexane), sodium trifluoroacetate, DowEX® marathon® MSC (H) ion-exchange resin were purchased from Sigma-Aldrich Chemicals Co. BO, DEAA, DMAA, diisopropylamine, chlorotrimethylsilane, and CH_2Cl_2 were distilled from CaH_2 and degassed by three freeze-pump-thaw cycles prior to use. Toluene and THF were distilled from sodium benzophenone ketyl. (*Z*)-1-Dimethylamino-1-trimethylsiloxy-1-propene ((*Z*)-DATP) [34], 1-ethoxyethyl glycidyl ether (EEGE) [41], *N*-(2-hydroxyethyl)-*N*-methylpropionamide [42], and (*Z*)-1-(*N*-methyl-*N*-(2-trimethylsiloxyethyl)amino)-1-trimethylsiloxy-1-propene ((*Z*)-HDATP) [42] were synthesized based on previous reports. Spectra/Por® 6 Membrane (MWCO: 1 000) was used for the dialysis. All other chemicals were purchased from available suppliers and used without purification.

4.2.2 Measurements

The ^1H (400 MHz) and ^{13}C NMR (100 MHz) spectra were recorded using a JEOL JNM-A400II and a JEOL-ECP400. The polymerization solution was prepared in an MBRAUN stainless steel glovebox equipped with a gas purification system (molecular sieves and copper catalyst) in a dry argon atmosphere (H_2O, O_2 < 1 ppm). The moisture and oxygen contents in the glovebox were monitored by an MB-MO-SE 1 and an MB-OX-SE 1, respectively. The degree of polymerization of each segment in the synthesized PDEAA-*b*-polyether ($DP_{PDEAA}/DP_{polyether}$) was determined by the ^{1}H NMR measurement.

Size exclusion chromatography (SEC) was performed at 40 °C using a Jasco high performance liquid chromatography (HPLC) system (PU-980 Intelligent HPLC pump, CO-965 column oven, RI-930 Intelligent RI detector, and Shodex DEGAS KT-16) equipped with a Shodex Asahipak GF-310 HQ column (linear, 7.6 mm × 300 mm; pore size, 20 nm; bead size, 5 μm; exclusion limit, 4 × 10^4) and a Shodex Asahipak GF-7 M HQ column (linear, 7.6 mm × 300 mm; pore size, 20 nm; bead size, 9 μm; exclusion limit, 4 × 10^7) in DMF containing lithium chloride (0.01 mol L^{-1}) at the flow rate of 0.4 mL min^{-1}. The number-average molecular weight (M_n) and polydispersity index (M_w/M_n) of the obtained polymers were determined by the RI based on PDMAA synthesized in our previous study [34] with the M_w (M_w/M_n)s of 5.77 × 10^4 g mol^{-1} (1.07), 2.54 × 10^4 g mol^{-1} (1.08), 1.32 × 10^4 g mol^{-1} (1.11), 9.60 × 10^3 g mol^{-1} (1.09), 7.10 × 10^3 g mol^{-1} (1.12), 5.10 × 10^3 g mol^{-1} (1.12), and 3.30 × 10^3 g mol^{-1} (1.16). The matrix-assisted laser desorption/ionization time-of-flight mass spectrometry (MALDI-TOF MS) measurements were performed using an Applied Biosystems Voyager-DE STR-H mass spectrometer with a 25 kV acceleration voltage. The positive ions were detected in the reflector mode (25 kV). A nitrogen laser (337 nm, 3 ns pulse width, 106–107 W cm^{-2}) operating at 3 Hz was used to produce the laser desorption, and 200 shots were summed. The spectra were externally calibrated using insulin (bovine) with a linear calibration. Samples for the MALDI-TOF MS were prepared by mixing the polymer (1.5 mg mL^{-1}, 10 μL), the matrix (*trans*-3-indoleacrylic acid, 10 mg mL^{-1}, 90 μL), and the cationizing agent (sodium trifluoroacetate, 10 mg mL^{-1}, 10 μL) in THF. The ultraviolet-visible (UV-vis) spectra were measured using a 10-mm path length cell by a Jasco V-550 spectrophotometer equipped with an EYELA NCB-1200 temperature controller and a Jasco ETC-505T temperature controller.

4.2.3 GTP of N,N-Diethylacrylamide (DEAA)

A typical procedure for the polymerization is as follows: to a solution of DEAA (0.513 g, 4.03 mmol) and (*Z*)-HDATP (12.3 μL, 40.0 μmol) in CH_2Cl_2 (7.40 mL), a stock solution of Tf_2NH in CH_2Cl_2 (40 μL, 0.8 μmol, 0.020 mol L^{-1}) was added at 20 °C to initiate the polymerization. After 2 h, the polymerization was quenched by adding a small amount of *t*BuOH/pyridine. The reaction mixture was stirred with the DowEX® marathon® MSC (H) ion-exchange resin in water and then dialyzed against methanol. The solvent was evaporated from the dialyzed solution to give a transparent solid. The solid was dissolved in water, then freeze-dried to give PDEAA-OH as a white solid. Yield, 496 mg (96 %).

M_n = 13.7 kg mol^{-1}, M_w/M_n = 1.05. The GTP of DEAA using (*Z*)-DATP was carried out by a similar procedure without treating the product with the ion-exchange resin.

4.2.4 Block Copolymerization of DEAA and N,N-Dimethylacrylamide (DMAA)

The first polymerization of DEAA in CH_2Cl_2 at 20 °C was carried out with $[DEAA]_0/[(Z)\text{-}DATP]_0/[Tf_2NH]_0$ = 50/1/0.02 for 1 h using the same procedure as the homopolymerization of DEAA. After an aliquot was taken from the reaction mixture to determine the conversion of DEAA and the M_n and M_w/M_n of the product, the block polymerization was carried out by adding 35 equivalents of DMAA. After 1 h, the polymerization was quenched by adding a small amount of *t*BuOH/pyridine. The product was purified by dialysis against methanol followed by freeze-drying from the aqueous solution to give poly(*N*,*N*-diethylacrylamide)-*block*-poly(*N*,*N*-dimethylacrylamide) (PDEAA-*b*-PDMAA) as a white solid. Yield = 520 mg (86 %). M_n = 8.05 kg mol^{-1}, M_w/M_n = 1.05 (PDEAA); M_n = 10.6 kg mol^{-1}, M_w/M_n = 1.06 (block copolymer). The block copolymerization using (*Z*)-HDATP was carried out using a similar procedure.

4.2.5 Synthesis of PDEAA-block-poly(ethylene oxide) ($PDEAA_{63}$-b-PEO_{44})

EO in THF (1.20 mol L^{-1}, 500 μL, 600 μmol) was added to a solution of $PDEAA_{63}$-OH (74.1 mg, 9.08 μmol, M_n = 8.16 kg mol^{-1}, M_w/M_n = 1.04, <DP> = 63) and *t*-Bu-P_4 (9.1 μL, 9.1 μmol, 1.0 mol^{-1} in *n*-hexane) in dry THF (87.9 μL) under an argon atmosphere. The reaction mixture was heated to 45 °C and stirred for 48 h. An excess amount of benzoic acid was added to the reaction mixture to quench the polymerization. The crude product was purified by dialysis against methanol followed by freeze-drying from the aqueous solution to give $PDEAA_{63}$-*b*-PEO_{44} as a white solid. Yield = 89.7 mg (90 %). DP_{PDEAA}/DP_{PEO} = 63/44.

4.2.6 Synthesis of PDEAA-block-poly(propylene oxide) ($PDEAA_{63}$-b-PPO_{44}), PDEAA-block-poly(1,2-butene oxide) ($PDEAA_{63}$-b-PBO_{41}), and PDEAA-block-poly (1-ethoxyethyl glycidyl ether) ($PDEAA_{63}$-b-$PEEGE_{40}$)

All the polymerizations were carried out using a similar procedure for the synthesis of $PDEAA_{63}$-*b*-PEO_{44} at 50 °C for 48 h.

The synthesis of the PDEAA$_{63}$-*b*-PPO$_{44}$ was carried out with PDEAA$_{63}$-OH (54.3 mg, 6.72 μmol), PO (19.7 μL, 282 μmol), THF (275 μL), and *t*-Bu-P$_4$ in *n*-hexane (1.0 mol L^{-1}, 6.72 μL, 6.72 μmol). The product was obtained as a white solid. Yield = 62.6 mg (92 %). DP$_{PDEAA}$/DP$_{PPO}$ = 63/44.

The synthesis of the PDEAA$_{63}$-*b*-PBO$_{41}$ was carried out with PDEAA$_{63}$-OH (63.9 mg, 7.83 μmol), BO (43.6 μL, 501 μmol), THF (450 μL), and *t*-Bu-P$_4$ in *n*-hexane (1.0 mol L^{-1}, 7.83 μL, 7.83 μmol). The product was obtained as a pale yellow solid. Yield = 83.2 mg (83 %). DP$_{PDEAA}$/DP$_{PBO}$ = 63/41.

The synthesis of the PDEAA$_{63}$-*b*-PEEGE$_{40}$ was carried out with PDEAA$_{63}$-OH (82.7 mg, 10.1 μmol), EEGE (96.8 μL, 649 μmol), THF (582 μL), and *t*-Bu-P$_4$ in *n*-hexane (1.0 mol L^{-1}, 10.1 μL, 10.1 μmol). The product was obtained as a pale brown solid. Yield = 145 mg (82 %). DP$_{PDEAA}$/DP$_{PEEGE}$ = 63/40.

4.2.7 *Synthesis of PDEAA-block-poly(glycidol) (PDEAA$_{63}$-b-PGD$_{40}$)*

Conc. HCl (250 μL) was added to a solution of PDEAA$_{63}$-*b*-PEEGE$_{40}$ (85 mg) in THF (1 mL). After stirring for 4.5 h at room temperature, a small amount of Amberlyst A21 was added to the reaction mixture for neutralization. After stirring for 1 h at room temperature, the crude product was purified by dialysis against methanol followed by freeze-drying from an aqueous solution to give PDMAA-*b*-PGD as a white solid. Yield = 47.3 mg (70 %). DP$_{DEAA}$/DP$_{GD}$ = 63/40.

4.2.8 *Turbidimetric Analysis*

An aqueous solution of a polymer sample (2.0 mg mL^{-1}) was prepared and transferred to a poly(methyl methacrylate) cell with a 1-cm path length. The cell was cooled to 8 °C under flowing nitrogen for 1 h. The absorbance at 500 nm of the aqueous solutions was recorded by a UV-vis spectrophotometer equipped with a temperature controller. The solution temperature was gradually increased at the heating rate of 1.0 °C min^{-1}.

4.3 Results and Discussion

4.3.1 *Group Transfer Polymerization (GTP) of N,N-Diethylacrylamide (DEAA)*

In order to synthesize PDEAA and PDEAA-OH, the Tf$_2$NH-promoted GTP of DEAA was carried out using (*Z*)-1-dimethylamino-1-trimethylsiloxy-1-propene ((*Z*)-DATP) and (*Z*)-1-(*N*-methyl-*N*-(2-trimethylsiloxyethyl)amino)-1-trimethylsiloxy-1-propene

Table 4.1 Results of Tf_2NH-promoted GTPs of DEAA using (*Z*)-DATP and (*Z*)-HDATP in CH_2Cl_2 at 20 °C[a]

Run	Initiator	$[DEAA]_0$ / $[initiator]_0$	Time (h)	M_n (kg mol^{-1}) Calcd.[b]	Obsd.[c]	M_w/M_n^c
1	(*Z*)-DATP	50	3	6.62	8.33	1.04
2	(*Z*)-DATP	100	3	13.1	14.6	1.05
3	(*Z*)-DATP	200	3	27.1	32.0	1.09
4	(*Z*)-HDATP	50	2	6.62	8.16	1.04
5	(*Z*)-HDATP	100	2	13.1	13.7	1.05
6	(*Z*)-HDATP	200	3	26.1	26.0	1.07

[a]$[DEAA]_0$ = 0.50 mol L^{-1}; $[initiator]_0/[Tf_2NH]_0$ = 1/0.02. [b]Calculated from $[DEAA]_0/([initiator]_0 - [Tf_2NH]_0)$ × (MW of DEAA = 127.18) + (MW of the initiator residue: (*Z*)-DATP = 101.13 (*Z*)-HDATP = 131.17). [c]Determined by SEC in DMF containing 0.01 mol L^{-1} of LiCl calibrated by the poly(*N*,*N*-dimethylacrylamide) (PDMAA) synthesized in our previous report [34]

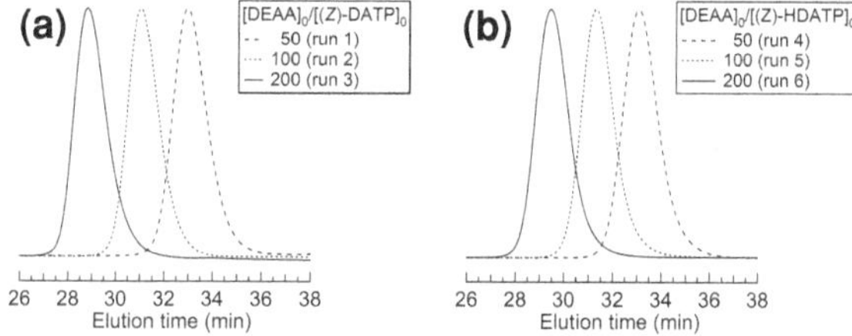

Fig. 4.1 The SEC traces of (**a**) the obtained PDEAAs (runs 1–3), and (**b**) the obtained PDEAA-OHs (runs 4–6) measured in DMF containing 0.01 mol L^{-1} of LiCl at 0.4 mL min^{-1} of the flow rate

((*Z*)-HDATP), amino silyl enolates, as the initiators. The design of (*Z*)-HDATP followed the structure of the silyl enolate used for the synthesis of hydroxyl end-functionalized poly(methyl methacrylate) by GTP [43, 44]. The GTP using (*Z*)-DATP was carried out in CH_2Cl_2 at 20 °C for 3 h under the condition of $[DEAA]_0/[(Z)\text{-DATP}]_0/[Tf_2NH]_0$ = 50/1/0.02, 100/1/0.02 and 200/1/0.02 (Table 4.1, runs 1–3). All the polymerizations homogeneously proceeded. The perfect conversion of DEAA was confirmed from the ^1H NMR spectra of the crude products. The SEC traces of the products showed a monomodal molecular weight distribution, as shown in Fig. 4.1a. The number-average molecular weight (M_n) and the polydispersity index (M_w/M_n) of the products were in a range of 8.33–32.0 and 1.05–1.09, which well agreed with their molecular weight ($M_{n,calcd.}$) calculated from the polymerization conditions. The GTP using (*Z*)-HDATP was carried out in CH_2Cl_2 at 20 °C for 2–3 h under the conditions of $[DEAA]_0/[(Z)\text{-HDATP}]_0/[Tf_2NH]_0$ = 50/1/0.02, 100/1/0.02, and 200/1/0.02 (Table 4.1, runs 4–6). All the polymerizations homogenously proceeded and DEAA was completely consumed. After the polymerization was quenched by adding *t*BuOH/pyridine, the crude product was stirred with a strong acid cation exchange resin in water to remove the trimethylsilyl group from the initiator residue. The M_n and M_w/M_n

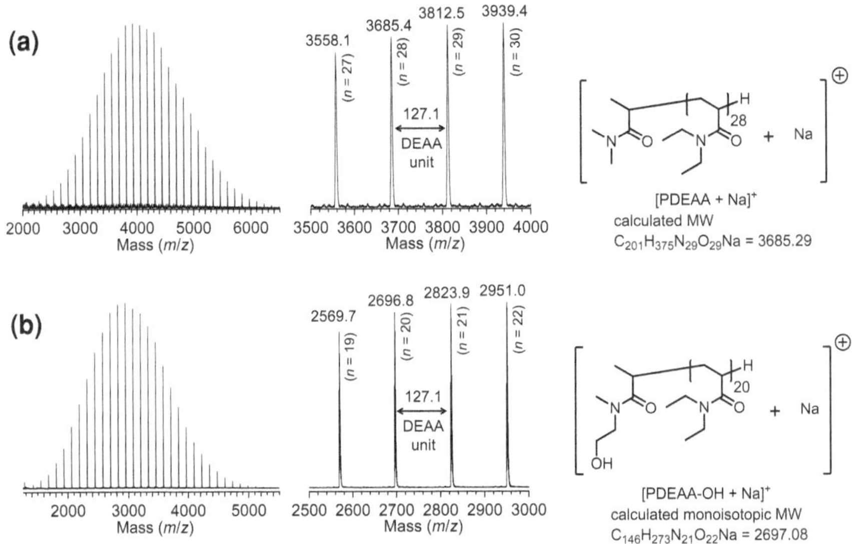

Fig. 4.2 MALDI-TOF MS spectra of (**a**) the obtained PDEAA ($[\text{DEAA}]_0/[(Z)\text{-DATP}]_0/[\text{Tf}_2\text{NH}]_0$ = 25/1/0.02, conversion >99.9 %, M_n = 4.90 kg mol^{-1}, M_w/M_n = 1.04), and (**b**) PDEAA-OH ($[\text{DEAA}]_0/[(Z)\text{-HDATP}]_0/[\text{Tf}_2\text{NH}]_0$ = 20/1/0.02, conversion >99.9 %, M_n = 3.72 kg mol^{-1}, M_w/M_n = 1.04)

of the products were in the range of 8.16–26.0 kg mol^{-1} and 1.04–1.07, which well agreed with their $M_{n,calcd}$. In addition, the SEC traces of all the products showed sharp and monomodal molecular weight distributions, as shown in Fig. 4.1b. Thus, the GTP of DEAA initiated by (*Z*)-DATP and (*Z*)-HDATP certainly proceeded and produced the products with predetermined M_n values and narrow M_w/M_n.

In order to confirm that an initiator residue was truly introduced into the obtained polymers, MALDI-TOF MS measurements were carried out for the products of the polymerization using (*Z*)-DATP and (*Z*)-HDATP. Only a series of molecular ion peaks was observed in the spectra of both products, as shown in Fig. 4.2. The difference in the *m*/*z* values among each molecular ion peak was just 127.1, which corresponded to the molecular weight of DEAA as a monomer unit. The *m*/*z* values of the molecular ion peaks in the spectra well agreed with the theoretical values for the adduct of the sodium ion and PDEAA bearing a hydrogen atom and 1-(*N*,*N*-dimethylcarbamoyl)ethyl group at each chain end (molecular formula: $C_{7n+5}H_{13n+11}O_{n+1}N_{n+1}Na$) or that of the sodium ion and PDEAA bearing a hydrogen atom and 1-(*N*-(2-hydroxyethyl)-*N*-methylaminocarbamoyl)ethyl group at each chain end (molecular formula: $C_{7n+6}H_{13n+13}O_{n+2}N_{n+1}Na$). Thus, the Tf_2NH-promoted GTP of DEAA was revealed to proceed without any side reactions, such as a back biting reaction [45, 46], to produce the desired PDEAA and PDEAA-OH. Importantly, the protected hydroxyl group of (*Z*)-HDATP was stable even under the conditions of the Tf_2NH-promoted GTP to realize the precise synthesis of PDEAA-OH by the GTP.

Table 4.2 Results of the sequential polymerization of DEAA and DMAA by Tf_2NH-promoted GTP using (*Z*)-DATP and (*Z*)-HDATP in CH_2Cl_2 at 20 °C[a]

Run	Initiator	Monomer	$[M]_0/[I]_0$	Time (h)	M_n (kg mol^{-1})		M_w/M_n[c]
					Calcd.[b]	Obsd.[c]	
7	(*Z*)-DATP	DEAA	50	1	6.59	8.05	1.05
		DMAA	35	1	10.1	10.6	1.06
8	(*Z*)-HDATP	DEAA	50	1	6.62	8.91	1.05
		DMAA	35	1	10.1	12.1	1.05

[a]$[DEAA]_0 = 0.50$ mol L^{-1}; $[initiator]_0/[Tf_2NH]_0 = 1/0.02$; monomer conversion >99.9 %. [b]Calculated from $[M]_0/([I]_0 - [Tf_2NH]_0) \times$ (MW of DEAA = 127.18 or DMAA = 99.13) + (MW of the initiator residue: (*Z*)-DATP = 101.13 (*Z*)-HDATP = 131.17; or $M_{n,calcd.}$ for the first polymer). [c]Determined by SEC in DMF containing 0.01 mol L^{-1} of LiCl calibrated by the PDMAA synthesized in our previous report [34]

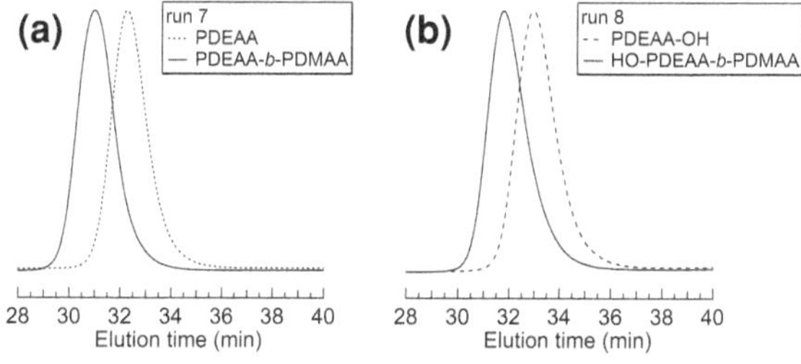

Fig. 4.3 The SEC traces of (**a**) the PDEAA and PDEAA-*b*-PDMAA obtained in run, and (**b**) the PDEAA-OH and HO-PDEAA-*b*-PDMAA obtained in run 8 measured in DMF containing 0.01 mol L^{-1} of LiCl at 0.4 mL min^{-1} of the flow rate

4.3.2 Synthesis of PDEAA-block-PDMAA by Sequential GTP

The block copolymerization of DEAA and DMAA was successful for the polymerizations using (*Z*)-DATP (Table 4.2, run 7). The first polymerization was carried out in CH_2Cl_2 at 20 °C under the conditions of $[DEAA]_0/[initiator]_0/[Tf_2NH]_0 = 50/1/0.02$ for 1 h. The DEAA was completely consumed to produce PDEAA with 8.05 kg mol^{-1} of M_n and 1.05 of M_w/M_n. The second polymerization was carried out for 1 h by adding 35 equivalents of DMAA just after the completion of the first polymerization. The M_n and M_w/M_n of the product were 10.6 kg mol^{-1} and 1.06, respectively. Both products had monomodal molecular weight distributions. The SEC traces of the PDEAA obtained from the first polymerization shifted to the region of high molecular weight after the second polymerization, as shown in Fig. 4.3a. Both segments of PDEAA and PDMAA were observed in the ^1H NMR spectrum of the product, as shown in Fig. 4.4, which indicated that the product is certainly PDEAA-*b*-PDMAA. This result proved that the silyl enolate at the chain end of the propagating PDEAA possessed a living nature in the Tf_2NH-promoted GTP. The block copolymerization of DEAA and DMAA was also successful even when (*Z*)-HDATP was used instead of (*Z*)-DATP

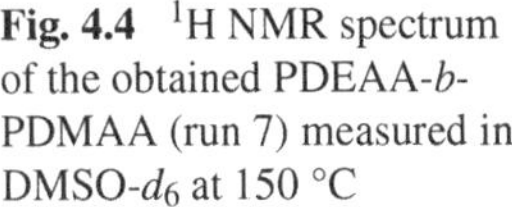

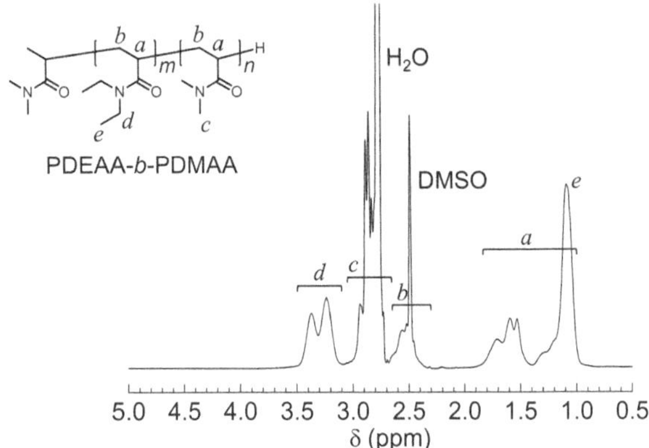

Fig. 4.4 ^{1}H NMR spectrum of the obtained PDEAA-*b*-PDMAA (run 7) measured in DMSO-d_6 at 150 °C

(run 8). Both of the products obtained after the first and second polymerizations had monomodal molecular weight distributions and the SEC traces of the PDEAA-OH obtained from the first polymerizations completely shifted to the region of high molecular weight after the second polymerization, as shown in Fig. 4.2d. Thus, it was proved that the sequential GTP using Tf_2NH and an amino silyl enolate is capable of synthesizing block copolymers consisting of a PDEAA segment and another polyacrylamide segment.

4.3.3 Synthesis of PDEAA-block-Polyethers

In order to demonstrate that the synthesized PDEAA-OH is applicable for the block copolymerization with various epoxides, PDEAA$_{63}$-OH (M_n = 8.16 kg mol^{-1}, M_w/M_n = 1.04, DP = 63, run 4) was employed as a macroinitiator for the *t*-Bu-P$_4$-catalyzed ROP of epoxides, as shown in Scheme 4.1. *t*-Bu-P$_4$ has been employed for the anionic ROP of epoxides, such as ethylene oxide (EO), propylene oxide (PO), and styrene oxide, initiated by various alcohols [47, 48]. EO and 1-ethoxyethyl glycidyl ether (EEGE), a protected glycidol, were chosen as monomers for the synthesis of the double-hydrophilic block copolymers [49, 50] based on the hydrophilicity of poly(ethylene oxide) (PEO) and poly(glycidol) (PGD). PO and 1,2-butene oxide (BO) were chosen for the synthesis of amphiphilic thermoresponsive block copolymers due to the hydrophobicity of the poly(propylene oxide) (PPO) and poly(1,2-butene oxide) (PBO). The polymerization was carried out under the conditions of [PO or BO]$_0$/[PDEAA$_{63}$-OH]$_0$/[*t*-Bu-P$_4$]$_0$ = 42/1/1 and [EO or EEGE]$_0$/[PDEAA$_{63}$-OH]$_0$/[*t*-Bu-P$_4$]$_0$ = 64/1/1 in THF at 45 or 50 °C for 48 h under an argon atmosphere. The polymerizations homogeneously proceeded and were quenched by adding an excess amount of benzoic acid to the reaction mixture. The epoxides were completely consumed in all the polymerizations as determined by ^{1}H NMR measurements. The signals due to the segments of PEO, PPO, PBO, and

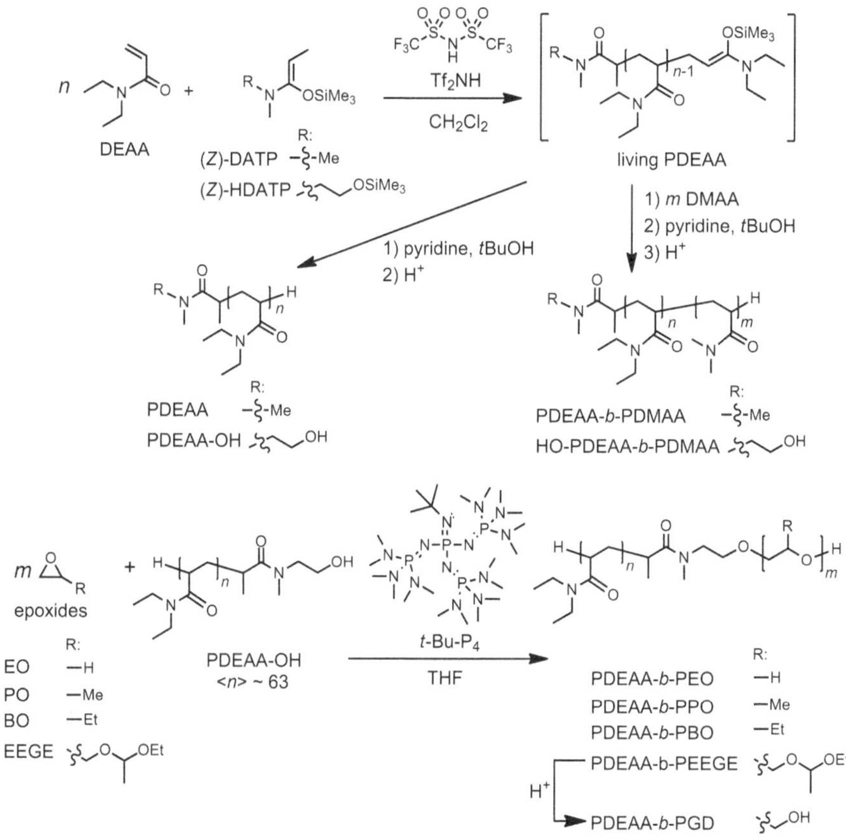

Scheme 4.1 Precise syntheses of PDEAA, PDEAA-OH, PDEAA-*b*-PDMAA, and PDEAA-*b*-polyether by Tf_2NH-promoted GTP and subsequent ring-opening polymerization (ROP) of epoxides using PDEAA-OH as the macroinitiator and *t*-BuP_4 as the catalyst

poly(1-ethoxyethyl glycidyl ether) (PEEGE) were observed in the ^{1}H NMR spectra of each product measured in $CDCl_3$, as shown in Figs. 4.5, 4.6, 4.7, and 4.8. The degree of polymerization of each segment ($DP_{PDEAA}/DP_{polyether}$) in the synthesized PDEAA-*block*-polyethers was determined from their ^{1}H NMR spectra as follows: 63/44 for $PDEAA_{63}$-*b*-PEO_{44} and $PDEAA_{63}$-*b*-PPO_{44}, 63/41 for $PDEAA_{63}$-*b*-PBO_{41}, and 63/40 for $PDEAA_{63}$-*b*-$PEEGE_{40}$. $PDEAA_{63}$-*b*-$PEEGE_{40}$ was treated with HCl in THF for 4.5 h at room temperature in order to convert the PEEGE segment into the PGD segment by deprotection of the 1-ethoxyethyl group. The successful deprotection was confirmed from the ^{1}H NMR spectrum of the product shown in Fig. 4.8b because the signal *h* due to the methine group of PEEGE originally observed at 4.7 ppm in Fig. 4.8a disappeared after the reaction. The value of DP_{PDEAA}/DP_{PGD} was 63/40 for $PDEAA_{63}$-*b*-PGD_{40}. These results proved that PDEAA-OH effectively acted as the macroinitiator for the polymerizations of epoxides to produce the various PDEAA-*block*-polyethers.

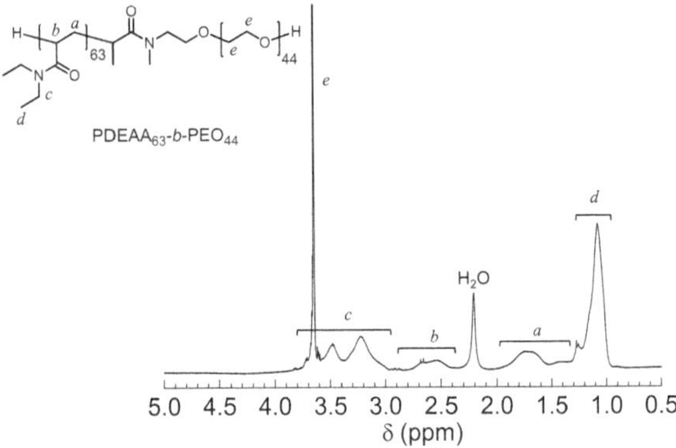

Fig. 4.5 ^{1}H NMR of the obtained PDEAA$_{63}$-*b*-PEO$_{44}$ measured in $CDCl_3$

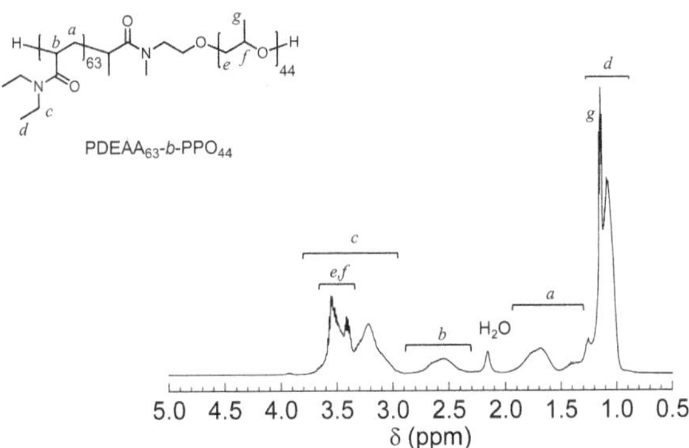

Fig. 4.6 ^{1}H NMR of the obtained PDEAA$_{63}$-*b*-PPO$_{44}$ measured in $CDCl_3$

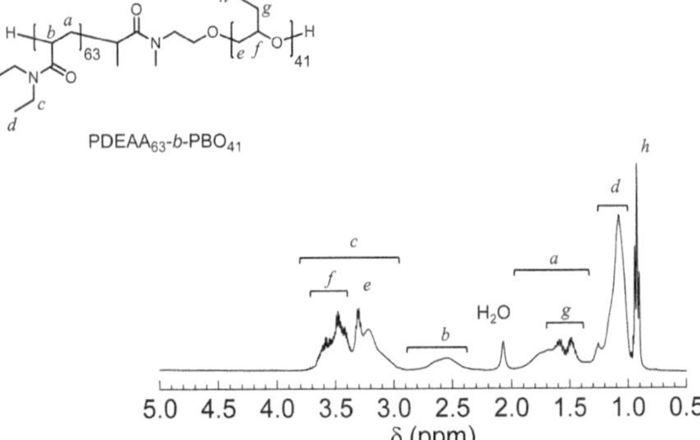

Fig. 4.7 ^{1}H NMR of the obtained PDEAA$_{63}$-*b*-PBO$_{41}$ measured in $CDCl_3$

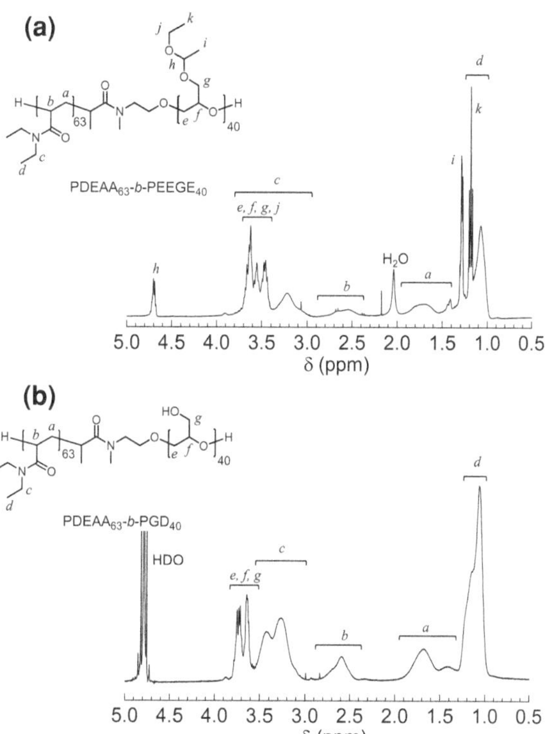

Fig. 4.8 ^{1}H NMR of the obtained (**a**) PDEAA$_{63}$-*b*-PEEGE$_{40}$ measured in $CDCl_3$, and (**b**) PDEAA$_{63}$-*b*-PGD$_{40}$ measured in D_2O

4.3.4 Thermoresponsive Properties of PDEAA-block-Polyethers

The thermoresponsive properties of the synthesized PDEAA-polyethers with different polyether segments were characterized by a turbidimetric analysis, as shown in Fig. 4.9. Every polymer showed the characteristic phase transition at a specific temperature depending on the properties of the polyether segment. The critical temperature of the phase transition (T_C) was defined as the temperature at which the absorbance of the aqueous solution of the sample begins increasing. Table 4.3 summarizes the determined T_Cs for each block copolymer. PDEAA$_{63}$-OH showed a T_C at 43.8 °C. The T_Cs of PDEAA$_{63}$-*b*-PPO$_{44}$ and PDEAA$_{63}$-*b*-PBO$_{41}$, which have a hydrophobic polyether segment, were 37.4 and 30.0 °C, respectively, and lower than the T_C of PDEAA$_{63}$-OH. The aqueous solution of PDEAA$_{63}$-*b*-PBO$_{41}$ was translucent even below its T_C probably due to the aggregation based on the hydrophobicity of the PBO segment, though that of PDEAA$_{63}$-*b*-PPO$_{41}$ was transparent below its T_C. On the other hand, the T_Cs of PDEAA$_{63}$-*b*-PGD$_{40}$, which has hydrophilic polyether segment, was 49.0 °C and higher than the T_C of PDEAA$_{63}$-OH. The aqueous solution of PDEAA$_{63}$-*b*-PGD$_{40}$ was transparent at room temperature based on the water-soluble PGD segment. Interestingly, PDEAA$_{63}$-*b*-PEO$_{44}$ showed

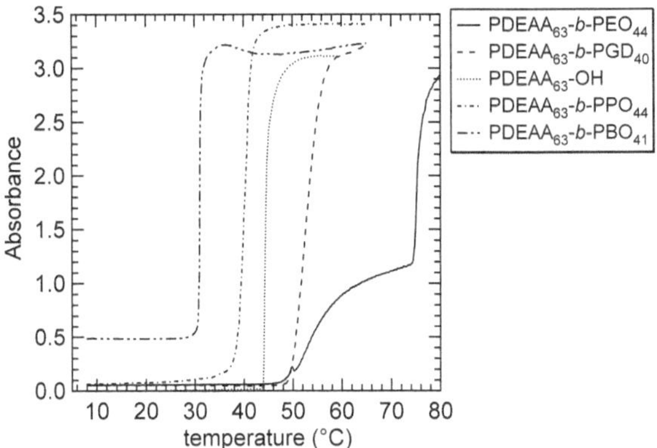

Fig. 4.9 Absorbance for aqueous solutions of $PDEAA_{63}$-OH, $PDEAA_{63}$-*b*-PEO_{44}, $PDEAA_{63}$-*b*-PPO_{44}, $PDEAA_{63}$-*b*-PBO_{41}, and $PDEAA_{63}$-*b*-PGD_{40} (2.0 mg mL^{-1}) measured by a light with 500 nm of wavelength at 1.0 °C min^{-1} of the heating rate

Table 4.3 Thermoresponsive properties of the synthesized PDEAA-*block*-polyethers

Polymer	T_C (°C)[a]
$PDEAA_{63}$-OH	43.8
$PDEAA_{63}$-*b*-PEO_{44}	48.4, 74.2
$PDEAA_{63}$-*b*-PPO_{44}	37.4
$PDEAA_{63}$-*b*-PBO_{41}	30.0
$PDEAA_{63}$-*b*-PGD_{40}	49.0

[a]Determined from the turbidimetric analysis

a two-stage phase transition at 48.4 and 74.2 °C. The solution became translucent after the first phase transition due to the solubility change in the PDEAA segment and eventual aggregation. The solution became turbid after the second phase transition plausibly due to the solubility change in the PEO segment. Thus, the T_C of the synthesized block copolymer was revealed to increase with the increasing hydrophilicity of the introduced polyether segment. In addition, the thermoresponsive properties and aggregation behavior of PDEAA in water was controllable by block copolymerization with epoxides. These results proved that the combination of the Tf_2NH-promoted GTP using (*Z*)-HDATP and the *t*-Bu-P_4-catalyzed anionic ROP of epoxides is a robust and versatile method for the preparation of various PDEAA-*block*-polyethers with different thermoresponsive properties.

4.4 Conclusions

The Tf_2NH-promoted GTP of DEAA has been achieved using (*Z*)-DATP, an amino silyl enolate, to produce the well-defined PDEAA. The hydroxyl end-functionalized PDEAA, i.e., PDEAA-OH, with the desired molecular weight and narrow M_w/M_n has been successfully synthesized from the same GTP using the amino silyl enolate bearing a protected hydroxyl group (*Z*)-HDATP, as the initiator. The sequential GTP using Tf_2NH and an amino silyl enolate is capable of synthesizing block copolymers consisting of a PDEAA segment and another polyacrylamide segment. In addition, the PDEAA-*block*-polyethers with different thermoresponsive properties were successfully produced by the polymerization of epoxides using PDEAA-OH as the macroinitiator. This study has provided a robust and versatile method for the design and precise synthesis of amphiphilic thermoresponsive block copolymers and double-hydrophilic block copolymers bearing a PDEAA segment.

References

1. Bütün V, Liu S, Weaver JVM, Bories-Azeau Y, Cai Y, Armes SP (2006) A brief review of schizophrenic block copolymers. React Funct Polym 66:157–165
2. Gil ES, Hudson SM (2004) Stimuli-reponsive polymers and their bioconjugates. Prog Polym Sci 29:1173–1222
3. Gaucher G, Dufresne M-H, Sant VP, Kang N, Maysinger D, Leroux J-C (2005) Block copolymer micelles: Preparation, characterization and application in drug delivery. J Controlled Release 109:169–188
4. Smith AE, Xu X, McCormick CL (2010) Stimuli-responsive amphiphilic (co)polymers via RAFT polymerization. Prog Polym Sci 35:45–93
5. Rapoport N (2007) Physical stimuli-responsive polymeric micelles for anti-cancer drug delivery. Prog Polym Sci 32:962–990
6. Blanazs A, Armes SP, Ryan AJ (2009) Self-assembled block copolymer aggregates: From micelles to vesicles and their biological applications. Macromol Rapid Commun 30:267–277
7. Dimitrov I, Trzebicka B, Müller AHE, Dworak A, Tsvetanov CB (2007) Thermosensitive water-soluble copolymers with doubly responsive reversibly interacting entities. Prog Polym Sci 32:1275–1343
8. Ito M, Ishizone T (2004) Synthesis of well-defined block copolymers containing poly (*N*-isopropylacrylamide) segment by anionic block copolymerization of *N*-methoxymethyl-*N*-isopropylacrylamide. Des Monomers Polym 7:11–24
9. André X, Zhang M, Müller AHE (2005) Thermo- and pH-responsive micelles of poly(acrylic acid)-*block*-poly(*N*,*N*-diethylacrylamide). Macromol Rapid Commun 26:558–563
10. André X, Benmohamed K, Yakimansky AV, Litvinenko GI, Müller AHE (2006) Anionic polymerization and block copolymerization of *N*,*N*-diethylacrylamide in the presence of triethylaluminum. kinetic investigation using in-line FT-NIR spectroscopy. Macromolecules 39:2773–2787
11. Vinogradova L, Fedorova L, Adler H-JP, Kuckling D, Seifert D, Tsvetanov CB (2005) Controlled anionic block copolymerization with *N*,*N*-dialkylacrylamide as a second block. Macromol Chem Phys 206:1126–1133
12. Angelopoulos SA, Tsitsilianis C (2006) Thermo-reversible hydrogels based on poly(*N*,*N*-diethylacrylamide)-*block*-poly(acrylic acid)-*block*-poly(*N*,*N*-diethylacrylamide) double hydrophilic triblock copolymer. Macromol Chem Phys 207:2188–2194
13. Hawker CJ, Bosman AW, Harth E (2001) New polymer synthesis by nitroxide mediated living radical polymerizations. Chem Rev 101:3661–3688
14. Rigolini J, Grassl B, Reynaud S, Billon L (2010) Microwave-assisted nitroxide-mediated polymerization for water-soluble homopolymers and block copolymers synthesis in homogeneous aqueous solution. J Polym Sci, Part A: Polym Chem 48:5775–5782
15. Matyjaszewski K, Xia J-H (2001) Atom transfer radical polymerization. Chem Rev 101:2921–2990
16. Ouchi M, Terashima T, Sawamoto M (2009) Transition metal-catalyzed living radical polymerization: Toward perfection in catalysis and precision polymer synthesis. Chem Rev 109:4963–5050

17. Xu Y, Shi L, Ma R, Zhang W, An Y, Zhu XX (2007) Synthesis and micellization of thermo- and pH-responsive block copolymer of poly(*N*-isopropylacrylamide)-*block*-poly(4-vinylpyridine). Polymer 48:1711–1717
18. Tang X, Liang X, Yang Q, Fan X, Shen Z, Zhou Q (2009) AB_2-type amphiphilic block copolymers composed of poly(ethylene glycol) and poly(*N*-isopropylacrylamide) via single-electron transfer living radical polymerization: Synthesis and characterization. J Polym Sci Part A 47:4420–4427
19. He X, Wu X, Gao C, Wang K, Lin S, Huang W, Xie M, Yan D (2011) Synthesis and self-assembly of a hydrophilic, thermo-responsive poly(ethylene oxide) monomethyl ether-*block*-poly(acrylic acid)-*block*-poly(*N*-isopropylacrylamide) copolymer to form micelles for drug delivery. React Funct Polym 71:544–552
20. Moad G, Rizzard E, Thang SH (2009) Living radical polymerization by the RAFT process – a second update. Aust J Chem 62:1402–1472
21. Semsarilar M, Perrier S (2010) 'Green' reversible addition-fragmentation chain-transfer (RAFT) polymerization. Nat Chem 2:811–820
22. Ge Z, Xie D, Chen D, Jiang X, Zhang Y, Liu H, Liu S (2007) Stimuli-responsive double hydrophilic block copolymer micelles with switchable catalytic activity. Macromolecules 40:3538–3546
23. Bian F, Xiang M, Yu W, Liu M (2008) Preparation and self-assembly behavior of thermosensitive polymeric micelles comprising poly(styrene-*b*-*N*,*N*-diethylacrylamide). J Appl Polym Sci 110:900–907
24. Maeda Y, Mochiduki H, Ikeda I (2004) Hydration changes during thermosensitive association of a block copolymer consisting of LCST and UCST blocks. Macromol Rapid Commun 25:1330–1334
25. Romão RIS, Beija M, Charreyre M-T, Farinha JPS, da Silva AMPSG, Martinho JMG (2010) Schizophrenic behavior of a thermoresponsive double hydrophilic diblock copolymer at the air-water Interface. Langmuir 26:1807–1815
26. Beija M, Fedorov A, Charreyre M-T, Martinho JMG (2010) Fluorescence anisotropy of hydrophobic probes in poly(*N*-decylacrylamide)-*block*-poly(*N*,*N*-diethylacrylamide) block copolymer aqueous solutions: Evidence of premicellar aggregates. J Phys Chem B 114:9977–9986
27. Prazeres TJV, Beija M, Charreyre M-T, Farinha JPS, Martinho JMG (2010) RAFT polymerization and self-assembly of thermoresponsive poly(*N*-decylacrylamide-*b*-*N*,*N*-diethylacrylamide) block copolymers bearing a phenanthrene fluorescent α-end group. Polymer 51:355–367
28. Mori H, Kato I, Saito S, Endo T (2010) Proline-based block copolymers displaying upper and lower critical solution temperatures. Macromolecules 43:1289–1298
29. Walther A, Millard P-E, Goldmann AS, Lovestead TM, Schacher F, Barner-Kowollik C, Müller AHE (2008) Bis-hydrophilic block terpolymers via RAFT polymerization: Toward dynamic micelles with tunable corona properties. Macromolecules 41:8608–8619
30. Rieger J, Grazon C, Charleux B, Alaimo D, Jérôme C (2009) Pegylated thermally responsive block copolymer micelles and nanogels via *In Situ* RAFT aqueous dispersion polymerization. J Polym Sci, Part A: Polym Chem 47:2373–2390
31. Lee H-N, Bai Z, Newell N, Lodge TP (2010) Micelle/Inverse micelle self-assembly of a PEO-PNIPAm block Copolymer in Ionic liquids with double thermoresponsivity. Macromolecules 43:9522–9528
32. Zhang X, Oddon M, Giani O, Monge S, Robin J-J (2010) Novel strategy for ROP of NCAs using thiols as Initiators: Synthesis of diblock copolymers based on polypeptides. Macromolecules 43:2654–2656
33. Klaikherd A, Nagamani C, Thayumanavan S (2009) Multi-stimuli sensitive amphiphilic block copolymer assemblies. J Am Chem Soc 131:4830–4838
34. Fuchise K, Sakai R, Satoh T, Sato S, Narumi A, Kawaguchi S, Kakuchi T (2010) Group transfer polymerization of *N*,*N*-dimethylacrylamide using novel efficient system consisting of dialkylamino silyl enol ether as an Initiator and strong brønsted acid as an organocatalyst. Macromolecules 43:5589–5594

35. Esswein B, Steidl NM, Möller M (1996) Anionic polymerization of oxirane in the presence of the polyiminophosphazene base *t*-Bu-P_4. Macromol Rapid Commun 17:143–148
36. Esswein B, Molenberg A, Möller M (1996) Use of polyiminophosphazene bases for ringopening polymerizations. Macromol Symp 107:331–340
37. Justynska J, Hordyjewicz Z, Schlaad H (2005) Toward a toolbox of functional block copolymers via free-radical addition of mercaptans. Polymer 46:12057–12064
38. Groenewolt M, Brezsinski T, Schlaad H, Antonietti M, Groh PW, Iván B (2005) Polyisobutylene-*block*-poly(ethylene oxide) for robust templating of highly ordered mesoporous materials. Adv Mater 17:1158–1162
39. Zhao J, Mountrichas G, Zhang G, Pispas S (2009) Amphiphilic polystyrene-*b*-poly (*p*-hydroxystyrene-*g*-ethylene oxide) block-graft copolymers via a combination of conventional and metal-free anionic polymerization. Macromolecules 42:8661–8668
40. Misaka H, Sakai R, Satoh T, Kakuchi T (2011) Synthesis of high molecular weight and end-functionalized poly(styrene oxide) by living ring-opening polymerization of styrene oxide using the alcohol/phosphazene base Initiating system. Macromolecules 44:9099–9107
41. Fitton AO, Hill J, Jane DE, Millar R (1987) Synthesis of simple oxetanes carrying reactive 2-substituents. Synthesis 1140–1142
42. Ohta S, Fuchise K, Satoh T, Kakuchi T (2012) Precise synthesis and end-functionalization of poly(*N*,*N*-diethylacrylamide) by group transfer polymerization using super bronsted acid. Polym Prepr (Am Chem Soc, Div Polym Chem) 53(1):111–112
43. Sogah DY, Hertler WR, Webster OW, Cohen GM (1987) Group transfer polymerization - polymerization of acrylic monomers. Macromolecules 20:1473–1488
44. Sogah DY, Webster OW (1983) Polym Sci Polym Lett Ed 21:927–931
45. Müller AHE (1990) Group transfer and anionic polymerization: A critical comparison. Makromol Chem Macromol Symp 32:87–104
46. Raynaud J, Gnanou Y, Taton D (2009) Group transfer polymerization of (meth)acrylic monomers catalyzed by *N*-heterocyclic carbenes and synthesis of all acrylic block copolymers: Evidence for an associative mechanism. Macromolecules 42:5996–6005
47. Boileau S, Illy N (2011) Activation in anionic polymerization: Why phosphazene bases are very exciting promoters. Prog Polym Sci 36:1132--1151
48. Misaka H, Sakai R, Satoh T, Kakuchi T (2011) Synthesis of high molecular weight and end-functionalized poly(styrene oxide) by living ring-opening polymerization of styrene oxide using the alcohol/phosphazene base Initiat-ing system. Macromolecules 44:9099--9107
49. Smith AE, Xua X-W McCormick CL (2010) Stimuli-responsive amphiphilic (co)polymers via RAFT polymerization. Prog Polym Sci 35:45--93
50. Ge Z-S, Liu S-Y (2009) Supramolecular self-assembly of nonlinear am-phiphilic and double hydrophilic block copolymers in aqueous solutions. Macromol Rapid Commun 30:1523--1532

Chapter 5
Conclusions

In this thesis, the author described the design and the precise synthesis of thermoresponsive polyacrylamides with novel primary structures. The effect of a hydrogen bonding unit on the thermoresponsive properties of poly(*N*-isopropylacrylamide) (PNIPAM) was clearly elucidated based on the precise synthesis of the polymer using the atom transfer radical polymerization (ATRP) and the copper-catalyzed azide-alkyne cycloaddition (CuAAC). In addition, the author achieved for the first time the precise synthesis of polyacrylamides by the group transfer polymerization (GTP). The GTP using bis(trifluoromethanesulfonyl)imide (Tf_2NH), a strong Brønsted acid, as the precatalyst and an amino silyl enolate as the initiator realized the precise synthesis of representative polyacrylamides, such as poly(*N*,*N*-dimethylacrylamide) (PDMAA) and poly(*N*,*N*-diethylacrylamide) (PDEAA), with well-defined structures including end-functionalized polymers and block copolymers, which provided a robust method to control the thermoresponsive properties of the polyacrylamides. Therefore, it is believed that the findings of this study should contribute to the development of the precise synthesis and rational macromolecular design for intelligent materials in the field of nanotechnology and bioengineering.

A summary of this thesis is as follows.

The ATRP and the CuAAC realized the precise synthesis of the PNIPAMs with the same number-average degree of polymerization and different urea groups at the chain end, which allowed comparing the effect of the introduced urea groups on the thermoresponsive properties of PNIPAM without effects from other structural factors. For the series of urea end-functionalized PNIPAMs, the intermolecular hydrogen bonding of the chain end urea group was revealed to work even in water, leading to the self-assembly of the polymer chains at a temperature below the cloud point. Such an antecedent aggregation made it possible to facilitate the phase transition process, resulting in a drastic decrease in the cloud point.

Tf_2NH, a strong Brønsted acid, was revealed to promote the GTP of DMAA. The initiating system consisting of (*Z*)-DATP and Tf_2NH was able to produce a PDMAA with a higher molecular weight than that of any polyacrylamides

K. Fuchise, *Design and Precise Synthesis of Thermoresponsive Polyacrylamides*, Springer Theses, DOI: 10.1007/978-4-431-55046-4_5, © Springer Japan 2014

previously synthesized by the GTP using other initiating systems. As a direct consequence of the homology in the structure of an initiator and the chain end of the propagating polymer, (*Z*)-DATP showed an excellent initiating efficiency and polymerization control as the initiator in comparison to MTS, the conventional initiator for the GTP. In addition, the living character of the GTP of DMAA using (*Z*)-DATP and Tf_2NH was strongly confirmed by a kinetic study, MALDI-TOF MS measurements, and a postpolymerization experiment. To the best of the author's knowledge, this result was the first demonstration of the living polymerization of an acrylamide monomer by GTP.

A facile and versatile method to synthesize thermoresponsive amphiphilic block copolymers and double-hydrophilic block copolymers bearing a PDEAA segment was developed using the GTP of DEAA using Tf_2NH and amino silyl enolates. The precise synthesis of the well-defined PDEAA was achieved by the Tf_2NH-promoted GTP using (*Z*)-DATP in a manner similar to the GTP of DMAA. The hydroxyl end-functionalized PDEAA, i.e., PDEAA-OH with a narrow molecular weight distribution, was successfully synthesized by the Tf_2NH-promoted GTP using (*Z*)-HDATP, the functional initiator with a protected hydroxyl group. The sequential GTP using Tf_2NH and an amino silyl enolate is capable of synthesizing block copolymers consisting of a PDEAA segment and another polyacrylamide segment. In addition, the anionic ring-opening polymerization of epoxides using PDEAA-OH as the macroinitiator and *t*-Bu-P_4 as the catalyst successfully produced various block copolymers of the PDEAA-*block*-polyethers with different thermoresponsive properties. The turbidimetric analysis proved that the thermoresponsive properties of the PDEAA-*b*-polyethers were successfully controlled as originally expected from the molecular design.

Curriculum Vitae

Keita Fuchise was born in Japan in 1985. He studied applied chemistry at Hokkaido University in Sapporo and carried out his doctoral studies under the supervision of Prof. Toyoji Kakuchi in the Graduate School of Engineering at Hokkaido University. He finished his PhD in 2012 by developing novel methods for the precise synthesis of polyacrylamides. He has joined the groups of Prof. Christopher Barner-Kowollik and Prof. Michael A. R. Meier at the Karlsruhe Institute of Technology (KIT), Germany, as a postdoc from March 2013.

K. Fuchise, *Design and Precise Synthesis of Thermoresponsive Polyacrylamides*,
Springer Theses, DOI: 10.1007/978-4-431-55046-4, © Springer Japan 2014

Zeitfracht Medien GmbH
Ferdinand-Jühlke-Straße 7
99095 Erfurt, Deutschland
produktsicherheit@kolibri360.de